BRAIDED STRUCTURES
AND COMPOSITES

Production, Properties,
Mechanics, and
Technical Applications

Composite Materials: Analysis and Design

Series Editor

Ever J. Barbero

PUBLISHED

BRAIDED STRUCTURES

AND

COMPOSITES

Production, Properties, Mechanics, and Technical Applications

Sohel Rana
Raul Fangueiro

CRC Press
Taylor & Francis Group
Boca Raton London New York

CRC Press is an imprint of the
Taylor & Francis Group, an **informa** business

CRC Press
Taylor & Francis Group
6000 Broken Sound Parkway NW, Suite 300
Boca Raton, FL 33487-2742

First issued in paperback 2018

© 2016 by Taylor & Francis Group, LLC
CRC Press is an imprint of Taylor & Francis Group, an Informa business

No claim to original U.S. Government works

ISBN-13: 978-1-4822-4500-4 (hbk)
ISBN-13: 978-1-138-79630-0 (pbk)

Visit the Taylor & Francis Web site at
http://www.taylorandfrancis.com

and the CRC Press Web site at
http://www.crcpress.com

Contents

SECTION I   Braided Structures, Properties, and Modeling

SECTION II   Application of Braided Structures and Composites

Series Preface

Since their commercial introduction half a century ago, composite materials are of widespread use in many industries. Application composite materials in aerospace, windmill blades, highway bridge retrofit, and so on, require designs that assure safe and reliable operation for 20 years or more. Using composite materials, virtually any property, such as stiffness, strength, thermal conductivity, and fire resistance, can be tailored to the user's needs by selecting the constituent material, their proportion and geometrical arrangement, and so on. In other words, the engineer is able to design the material concurrently with the structure. Also, modes of failure are much more complex in composites than in classical materials. Such demands for performance, safety, and reliability require that engineers consider a variety of phenomena during the design. Therefore, the aim of the *Composite Materials: Analysis and Design* book series is to bring to the design engineer a collection of works written by experts on every aspect of composite materials that is relevant to their design.

Variety and sophistication of material systems and processing techniques have grown exponentially in response to an ever-increasing number and type of applications. Given the variety of composite materials available as well as their continuous change and improvement, understanding of composite materials is by no means complete. Therefore, this book series serves not only the practicing engineer but also the researcher and student who are looking to advance the state of the art in understanding material and structural response and developing new engineering tools for modeling and predicting such responses.

Thus, the series is focused on bringing to the public existing and developing knowledge about the material–property relationships, processing–property relationships, and structural response of composite materials and structures. The scope of the series includes analytical, experimental, and numerical methods that have a clear impact on the design of composite structures.

Preface

Braiding is a very old textile manufacturing technique and has been traditionally used for many years to produce ropes, shoe laces, cables, and so on. Recently, this technology is getting special attention in various technical fields, such as medical, aerospace industries, civil engineering, transportation, and others, due to its ability to produce structures that can fulfill the demands imposed by these technical sectors. Each of these applications needs special structures with special functionalities. Therefore, the design of these technical products requires a proper understanding of the process, mechanics, materials, structures, and various parameters. Some of these important aspects of braiding technology have been discussed separately in books and mainly in research articles. However, the main aim of writing this book is to discuss all of this important information together, focusing on the practical applications of braiding technology. This effort will be very fruitful for the engineers and scientists working in various industries to design their products or for the students and researchers to know this technology starting from the concept up to the product designing and applications.

This book presents various applications of braiding technology, which are already commercialized or in research stage and the recent advancements in terms of materials, processes, and structures to achieve the required functionalities. Each application sector has been dealt with in detail in separate chapters and written by the leading experts from each field, identifying the requirements and discussing the type of material, process, structure, and design strategy to fulfill these requirements.

One important issue dealt with in this book is the mechanics of the braiding process. It is extremely necessary to understand the mechanics to identify the main parameters influencing the braiding process and produced structures and to control these parameters to achieve the desired material characteristics. This book also discusses the modeling of structure and various properties of braided structures and available techniques and software tools used for that purpose. This discussion will be highly helpful for designing various types of braided products with required properties.

One complete chapter of the book has been written on braided composites. This new type of composite materials is becoming very attractive due to its benefits over conventional materials and composites. This chapter discusses the types of braided composites, their production process and new developments, properties, and applications. Detailed analysis of structure and braided composites is also presented in a separate chapter of the book.

The last chapter of the book highlights the importance of recently developed multiscale braided structures and composites. The understanding of these new materials can help to develop multifunctional products with various interesting features using braiding technology.

The specialty of this book is that it discusses the basic principles, processes, mechanics, analysis, product designing, and applications, as well as recent developments in braiding technology in terms of materials, processes, structures, and product designing. Therefore, this book has been written targeting students, teachers, researchers, engineers, and scientists working in various industries.

We express our sincere thanks and gratitude to all contributors who participated in the various chapters of this book, for the efforts and valuable contributions. Sincere thanks are also due to our colleagues from the Fibrous Materials Research Group, School of Engineering, University of Minho, for their kind help and support. We believe that this book will be an important reference to learn about the basic concepts related to braiding process and structures, to know the various applications of these materials, and also, to learn about the advanced analytical techniques required for product designing purposes. We have no doubt that a wide range of readers including students, teachers, engineers, researchers, and scientists will benefit from the discussions about braiding process, materials, products, applications, and modeling, presented in various chapters of this book.

Sohel Rana and Raul Fangueiro

Editor Bios

Dr. Sohel Rana is currently the scientific manager at Fibrous Materials Research Group, School of Engineering, University of Minho, Portugal. He obtained his bachelor's degree in textile technology from the University of Calcutta, India and master's degree and PhD in fiber science and technology from Indian Institute of Technology (IIT, Delhi), India. His current research areas are advanced fibrous and composite materials, natural fibers, nanocomposites, electrospinning, multifunctional and biocomposite materials, and so on. He has authored one book, more than 10 book chapters in edited books, two keynote articles, and over 80 publications in various refereed journals and international conferences. Currently, he is editing four books on composites and nanomaterials for leading publishers. He also participates on the editorial board of several scientific journals and is a potential reviewer for numerous scientific journals including *Composite Science and Technology*, *Composites Part A*, *Composite Interfaces*, *Powder Technology*, *Journal of Nanomaterials*, *Journal of Applied Polymer Science*, and so on.

Raul Fangueiro is currently a professor and researcher at the School of Engineering, University of Minho, holding a PhD in textile engineering. He is the coordinator of the Fibrous Materials Research Group, leading several research projects on fibrous and composite materials, mainly for building, architectural, and health-care applications. He has published more than 120 scientific articles in international journals and has also presented more than 350 communications in international conferences. He is the author and co-author of 14 technical books, 12 patents, and is supervising various PhD and post-doc scientific works. He is an expert of the European Textile Platform and member of the editorial board of several international scientific journals. He is the coordinator of the FIBRENAMICS International Platform (www.fibrenamics.com) including 190 partners, developing promotion, dissemination, technology transfer, and research activities on fiber-based advanced materials.

Contributors

R. Alagirusamy
Department of Textile Technology
Indian Institute of Technology Delhi
 (IIT Delhi)
Hauz Khas, New Delhi, India

Kadir Bilisik
Department of Textile Engineering
Faculty of Engineering
Erciyes University
Talas-Kayseri, Turkey

Nedim Erman Bilisik
Department of Electronic
 Engineering
Faculty of Engineering
Istanbul Kultur University
Bakirkoy-Istanbul, Turkey

Fernando Cunha
Department of Civil Engineering
University of Minho
Guimarães, Portugal

Guodong Fang
Centre for Composite Materials and
 Structures
Harbin Institute of Technology
Harbin, China

Raul Fangueiro
Department of Civil Engineering
University of Minho
Guimarães, Portugal

Aron Gabor
Laboratory of Composite Materials for
 Constructions
Claude Bernard Lyon 1 University
Villeurbanne, France

Bohong Gu
College of Textile
Donghua University
Shanghai, China

Nesrin Sahbaz Karaduman
Akdagmadeni Vocational High School
Bozok University
Akdagmadeni-Yozgat, Turkey

Jun Liang
Centre for Composite Materials and
 Structures
Harbin Institute of Technology
Harbin, China

Daniel Oliveira
Department of Civil Engineering
University of Minho
Guimarães, Portugal

Naveen V. Padaki
Central Silk Technological Research
 Institute
Central Silk Board, Ministry of
 Textiles,
Government of India
Bangalore, India

Sohel Rana
Fibrous Materials Research Group
School of Engineering
University of Minho
Guimarães, Portugal

Claudio Scarponi
Department of Mechanical and
 Aerospace Engineering
Sapienza University of Rome
Rome, Italy

Graça Vasconcelos
Department of Civil Engineering
University of Minho
Guimarães, Portugal

Wen Zhong
Department of Biosystem Engineering
University of Manitoba
Winnipeg, Manitoba, Canada

Section I

Braided Structures, Properties, and Modeling

1 Introduction to Braiding

R. Alagirusamy and Naveen V. Padaki

CONTENTS

ABSTRACT

Braiding is an important textile process that involves the least preparatory processes due to which the fabric preparation cost is much cheaper. It is also a flexible process for producing various profiled fabrics, which is of high significance in the field of three-dimensional (3D) preform production for composites. In this chapter, introduction to braiding technology, its advantages, and the history of braids have been explained. Comparison of braiding with other fabric formation techniques has been carried out. Classification of braided structures with respect to method of intertwining, shape, and applications has been detailed. Machineries used for two-dimensional (2D) and 3D braided structure preparation have been explained. The 3D braided structures prepared by Cartesian four-step and multistep processes along with hexagonal braiding technology offers versatility for the braids to achieve complex shapes and meet the design requirements of many technical applications.

1.1   WHAT IS BRAIDING?

Braid is a minor but distinctive form of textile fabric. The *Encyclopaedia Britannica* defines "braiding" as "in textiles, machine or hand method of interlacing three or more yarns or bias-cut cloth strips in such a way that they cross one another and are laid together in diagonal formation, forming a narrow strip of flat or tubular fabric" and also identifies "braid" as fabrics "made by interlacing three or more yarns or fabric strips, forming a flat or tubular narrow fabric." Braiding is a textile manufacturing process that is used to manufacture fabrics through intertwining of yarns in a diagonally overlapping manner. A comparison of textile fabric manufacturing processes [1] is presented in Table 1.1. Weaving is a process of fabric formation by interlacing of two sets of yarns (warp and weft) perpendicular to each other, whereas knitting is a process of fabric formation by interloping of yarns (drawing loops through previously created loops of yarn). Nonwoven fabrics are formed by interlocking the fibrous assembly through mechanical, chemical, or thermal means. Braiding is a process of fabric manufacture intertwining three or more yarns in a diagonally overlapping way so that no two yarns are twisted around one another. Plaiting of human hair is a very good example of braiding, which has uniform shape for personal adornment. Braided textiles have been used for centuries for applications such as ropes, cordages, and laces. Other uses include parachute cords, cord shock absorbers, fishing lines, wicking, packing, clothes lines, driving bands, flexible hose coverings, insulated sleeving, tapes and bindings, ties and girdles, coat, and dress and lingerie trimmings and tubes for rubberizing, and recently as reinforcements in composite materials.

Braiding was originally a manual process, but now it has evolved into an industrial fabric manufacturing process with the development of modern braiding machines and applications of braids in a variety of technical textile sectors. Braiding is one of the most cost-effective fabric manufacturing processes, which offers an excellent platform for the creation of complex and near net shape 3D structures. The concept of adding strength in the through-the-thickness direction of fabric to create 3D structures through the incorporation of fibers and yarns in the z direction by stitching, weaving, knitting, and braiding has been considered to reduce fabric manufacturing cost, increase through-the-thickness mechanical properties, and improve impact damage tolerance [2]. Figure 1.1 presents the classification of 3D textile structures based on different manufacturing processes.

TABLE 1.1
Comparison of Fabric Formation Techniques

Parameter	Weaving	Knitting	Nonwoven	Braiding
Basic direction of yarn introduction	Two sets of yarn (0° warp and 90° weft)	One set of yarn (0° warp or 90° weft)	Randomly arranged fibers	One set of yarn (along machine direction)
Fabric formation technique	Interlacing	Interlooping	Interlocking	Intertwining

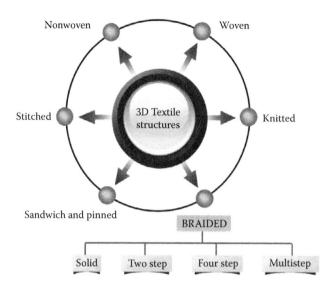

FIGURE 1.1   Three-dimensional (3D) textile structures.

1.2   HISTORY OF BRAIDING

Handmade braids have been discovered at Peruvian burial sites dating as far back as 8600–5780 BC [3] in the forms of ropes, looped bags, and twined fabrics [4]; these braids are likely to have been created far before their woven counterparts due to their simple formation and high functional value. An anthropological finding of braided cord impressions in ceramic pottery has been observed that dates back to the fourth millennium BC [5]. Rugs created by braiding long strips of fabric together have been created long before the time of Christ in virtually every country all over the world [6]. Furthermore, braids are known to have been developed and used in the Egyptian, Huari, Chavin, and Japanese cultures for both functional and decorative uses (flat braids being the oldest form of braids). Though there are many varieties in culturally constructed braids, those of the Peruvian and Japanese are the most widely studied and mimicked [4].

The Japanese art of silk braiding is known as "kumihimo," *kumi* translating to plait or braid and *himo* to string, cord, or braid. Many kumihimo braids are found in museums and private collections around the world. The earliest of kumihimo braids date back to 8000 BC and were thought to have been "loop manipulated" similar to "chain-stitched" knitted structure. The braiding apparatuses famously used in kumihimo have not been recorded to be used until 645–784 AD, although most believe they were in use before that time [6]. These braids were especially important in times of war when Owen [4] explains that 800–1000 ft of kumihimo braids were used to cover and lace together a suit of armor. Kumihimo braids were (and still are) used to wrap sword handles, hold together obi belts, and adorn hair and belts. Additionally, certain patterns and colors were reserved for members of society holding a certain title or rank.

Sling braids are of special interest and have been carefully studied and documented for their color patterns and functionality. The sling braid served as a weapon in which a pouch created in the middle of the braid held a projectile, which would be released through a swinging motion of the braid. Sling braids were dominantly found in the Incan and Peruvian cultures but have also been documented in Africa, India, Tibet, and the Far East and are known to have been used until the seventeenth century. Though the formation of the actual braid remained somewhat homogenous, the intricate patterns on these weapons varied greatly. Peruvian sling braids periodically adopted a technique in which the elements that form an inner core braid were brought to the outside of the top braid sheath to design more interesting structures. Aside from Japanese and Peruvian braiding, other forms of braiding documented throughout history include hair plaiting and maypole dancing. Additionally, braids with looped elements were unearthed alongside the relics from medieval churches in Finland and Sweden [7].

1.3 TYPES OF BRAIDING AND BRAIDED STRUCTURES

Braids can be classified based on various methods such as yarn intertwining, orientation of fibers/yarns in the braided structure, shape of the braided fabric and purpose of its use, as detailed here:

Regular braids: Regular braid constructions are 2/2 patterns of yarns intertwining as represented in Figure 1.2a, and the braided structure is similar to the 2/2 twill woven fabric structure.

Diamond braids: The 1/1 intertwining of yarns during braiding results in the diamond braided structure as represented in Figure 1.2b, which is similar to plain woven fabric structure.

Basket braids: These are 2 × 2 braided structures, which are similar to a 2 × 2 box weave with two yarns crossing over and under each other, as shown in Figure 1.2c.

Hercules braids: The 3/3 intertwining of yarns during braiding produces the Hercules braided structure as represented in Figure 1.2d, which is similar to the 3/3 twill woven fabric structure.

Biaxial and triaxial braids: Depending on the orientation and arrangement of input yarns during intertwining, braided structures can be categorized into

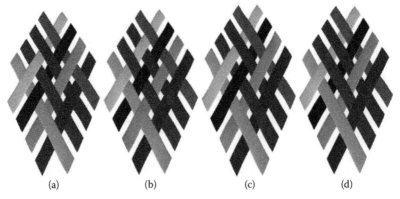

(a)   (b)   (c)   (d)

FIGURE 1.2 Braided structures with different intertwinings: (a) regular, (b) diamond, (c) basket, and (d) Hercules braids.

biaxial and triaxial structures. Braiding process usually results in biaxial structures, where two sets of yarns moving in opposite directions are intertwined at an angle. Incorporation of in-laid yarns along the machine direction or at cross direction during braiding results in triaxial structures, as represented in Figure 1.3.

Tubular, flat, square, and solid braids: Tubular braids are made on circular braiding machines, and they are composed of two sets of yarns along the axial direction of braided structure in which the first set spirals continuously in a clockwise direction while intertwining with the second set, which spirals in an anticlockwise direction. The counterclockwise and clockwise routes cause the sets of yarns to intersect in a full circular path, with each set of yarns moving simultaneously, resulting in a tubular structure. Each set of yarns, traveling either counterclockwise or clockwise, will never intersect with those sequential yarns moving in the same direction. Tubular braiding produces braids with rotational symmetry.

Flat braids, as the name implies, are ribbon-shaped braided structures. Unlike the spiral type of yarn path in the tubular braids, flat braids have clockwise and counterclockwise intertwining sets of yarns which reverse at the edges and travel back in the opposite direction, forming a selvedge braided structure, as shown in the Figure 1.4.

The square braid is a variation of a tubular braid wherein yarn groups travel in clockwise and counterclockwise directions while interlacing with yarns of the opposite group and maintain a hollow center. The square shape is created when the braid yarns are maneuvered into a square formation created by the same formation of the machinery's horn gears [10], as shown in Figure 1.5 [11].

Solid braids are structures with a plurality of interlocked braided layers. Selective interlocking of braided layers can be achieved through controlled movement of carrier yarns during braiding by which complex 3D shapes can be obtained [12]. The most common solid braided structures like 2 diagonal, 3 diagonal, 4 diagonal, spiral, and round braided structures are represented in Figure 1.5.

Biaxial braid Triaxial braid

FIGURE 1.3 Biaxial and triaxial braided structures. (From Braley, M., and M. Dingeldein, Advancements in braided materials technology, A&P Technology, Inc., *46th International SAMPE Symposium*, 2445–2454, 2001.) [8].

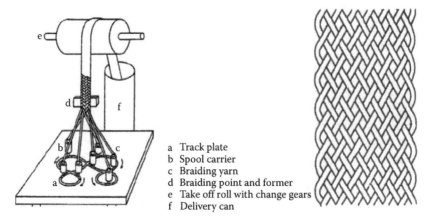

FIGURE 1.4 Schematic of the flat braiding machine and flat braided structure. (From Branscomb et al., *J. Eng. Fiber. Fabr.*, 8(2), 11–24, 2013.) [9].

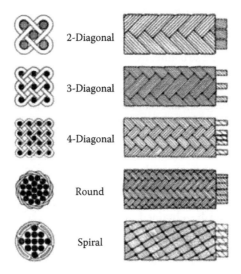

FIGURE 1.5 Solid braided structures. (From Tada, M., *Proceeding of 4th International Symposium on Textile Composites*, Kyoto, Japan, O-41-1, Retrieved from http://www .kumihimo-society.org/kumihimo/Key14.html, 1998.) [11].

Fancy braids: Fancy braids are structures created by braiding that are irregularly shaped or that are not simply flat or tubular in construction. Fancy braids are used as garment accessories, trimmings, and fashion ornamentations. The types of fancy braids are numerous and can include variations in structure, color, interlacement type, and so on. Common fancy braids are illustrated in Figure 1.6 [3], in which argyle (Figure 1.6a) braids display overlapping diamonds design; thin stripes along the axis of the braided structure are characteristics of pinstripe braids (Figure 1.6b); and horizontal zigzag striped braids (Figure 1.6c), as the name implies, contain horizontal stripes.

FIGURE 1.6 Fancy braided structures: (a) argyle, (b) pinstripe, and (c) zigzag striped braids. (From Bicking, A.M., Explorations in fancy braid creation through the use of industrial machinery, North Carolina State University, Raleigh, North Carolina, 2011.) [3].

1.4   TYPES OF FIBERS

Braided structures are fabrics that are relatively easy to produce without involving much yarn preparatory processes. That is the main reason why braids are one of the most cost-effective fabrics to manufacture. Also, yarn path in the braided structure is simple and straight enough so that even the most difficult and least flexible yarns can be converted into braided structure. Starting from natural fibers such as cotton, silk, wool, and linen, regenerated and synthetic fibers of all compositions find it easy to be braided. Even metallic fibers such as steel are converted into rope form using braiding technology (Figure 1.7) for civil infrastructure applications [13].

1.5   BRAIDING MACHINERY

Braiding machines in terms of production of 2D and 3D braids are detailed in the following subsections.

1.5.1   TWO-DIMENSIONAL BRAIDING

Thomas Walford patented the first braiding machine in 1748 for "laying or intermixing of threads, cords, or thongs of different kinds, commonly called plaiting" [1]. Later, in 1767, Bockmuhl of Germany introduced a simple braiding machine. In 1823, John Heathcoat introduced improved modifications to the braiding machinery [14]. These early machines were designed for the production of simple laces, tapes, and cords designed for use in garments and shoelaces. Most of these machines created 2D structures. The important 2D braiding machines are based on maypole braiding techniques, which are detailed in Sections 1.5.1.1 through 1.5.1.3.

FIGURE 1.7   Braided steel wire rope. (From Mark F, *Stainless steel XC expandable braided sleeving Techflex*, Flexible Technologies Limited, Auckland, 2014, retrieved from http://www.techflex.co.nz/stainless-steel-xc.html.) [13].

1.5.1.1   Conventional Maypole Braiders

Maypole is a term that specifies a folk dance performed by Europeans during the month of May around a pole. Maypole braiding machine derives its name from the dance, as the braiding is achieved in the machine by the movement of yarn carriers similar to the steps of the said dance. "Maypole" type of braiding machine (Figure 1.8) comprises a horizontal deck or track plate provided with grooves, slots, or tracks extending about the machine axis in which the bobbin carriers are guided around the axis of the machine [9]. There are usually two guideways in the track plate (clockwise and anticlockwise) leading in a tortuous or sinuous course around the central axis of the machine, crossing each other at regular intervals. Two sets of bobbin carriers are traversed in opposite directions in these tracks, to cause the carriers to pass inwardly and outwardly around each other to intertwine their yarns. A yarn extends upwardly under suitable tension provided by the tensioners located in the carriers, from the bobbin on each of the carriers to a point adjacent to the vertical axis of the braiding machine, and due to the intertwining of these yarns braided structure is formed about the said axis.

1.5.1.2   Modern Maypole Braiders

Improvisations in the maypole type of braiding machinery have resulted in a variety of mechanisms being introduced to increase the speed and efficiency of braiding machines [15]. There have been many patents for the designs of braiding machines or braiding mechanisms. J. Lundgren patented a braiding machine design with two additional sets of thread supplying devices in 1903 [16]. Since the two sets of devices are stationary, the braiding products have additional straight longitudinal threads interwoven with the braided threads. In 1912, S. W. Wardwell invented his famous braiding machine, which was later called the Wardwell braider [17,18]. The machine drives one set of carriers in one direction by a rotating plate and the other package of carriers in the opposite direction by two driving bars for each carrier. The driving bars were set in a cam profile. This improvement simplified the old high-speed braiders by replacing the plurality of gears with the same cam mechanism. The machine speed and efficiency were increased. Some later patents were based on similar braiding machine designs [19,20]. Fisher invented an apparatus and a method to control speed, especially, operating speed, which improved the efficiency of the braiding machine [21].

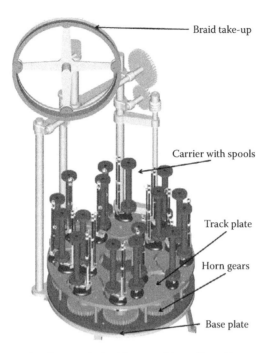

FIGURE 1.8   Schematic of maypole braiding machine. (From Branscomb et al., *J. Eng. Fiber. Fabr.*, 8(2), 11–24, 2013.) [9].

1.5.1.3   High-Speed Braiders

A rapid braider or rotary high-speed braiding machine [9] with the opposing bobbin carriers traveling in a circular rather than a sinusoidal motion and only the yarns following sinusoidal paths was patented in 1890 [22]. Rotary braiders represent an evolution of the maypole braiding machines that employ significant modifications and greatly improve braiding formation speed [23,24]. In traditional rotary braiding machines, counterrotating bobbin carriers move in perfect circles around banked edges of a bell so that the outer yarns interlace over and under inner yarns with the aid of deflector plates. In this way, the yarns with low mass and inertia undulate, rather than the bobbin carriers.

1.5.2   THREE-DIMENSIONAL BRAIDING

Three-dimensional braids are formed on two basic types of machines. These are the horn gear and Cartesian machines, which differ only in their method of yarn carrier displacement. While the horn gear-type machines offer improved braid speed over the Cartesian machines, the Cartesian machines offer compact machine size, comparatively low development cost, and braid architectural versatility. Hexagonal braiders are the recent addition to the 3D braiding processes, which also have been deliberated in this chapter.

1.5.2.1 Circular and Solid Braiders

Circular and solid braiders are horn gear type of machines, as shown in Figure 1.9, wherein the bobbins move continuously without any stoppage. They move on the track plate through the complete structure, around the standing ends, so that the movements of bobbins are faster compared to the Cartesian (four-step or multistep) braiding process. The bobbins can move only in two directions, so the horn gear process is also called two-step braiding process [11].

Track plate geometries other than a cylindrical locus for horn geared braiders enable the construction of complex, even solid shapes with included holes and bifurcations, with or without the insertion of axial yarns. Modified track plates allow the construction of noncircular cross section braids with reinforcement through the thickness. Transverse yarn interaction may occur throughout the thickness of the braid, resulting in solid structures with excellent toughness and delamination resistance [9].

1.5.2.2 Cartesian Braiding

Cartesian braiding is also known as track and column braiding, in which the groups of yarns traverse in cycles of distinct steps of horizontal and vertical movements. Typically, Cartesian braiders are used to prepare 3D braids by four-step method. Recent advancement in Cartesian braiding has witnessed multistep braiding, which provides innumerable varieties of 3D shapes to be accomplished through this method of braiding.

1.5.2.2.1 Four-Step Cartesian Braiders

A four-step process has four distinct cycles of horizontal or vertical movements in which the groups of yarns travel in; each movement control has a predetermined path and distance for each of the yarn groups. In each step, the bobbins move to the neighboring crossing point in two ways and stop for a specific interval of time. Basic arrangement of the braiding field is obtained after a minimum of four steps [25]. After these four motions are complete, one machine cycle is finished and beat up

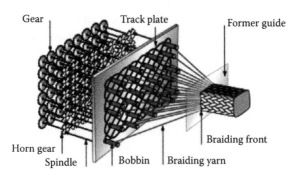

FIGURE 1.9 Schematic of the solid braid preparation through horn gear process. (From Tada, M., *Proceeding of 4th International Symposium on Textile Composites*, Kyoto, Japan, O-41-1, Retrieved from http://www.kumihimo-society.org/kumihimo/Key14.html, 1998.) [11].

is necessary to drive the yarns into the structure at the completion of each cycle, as shown in Figure 1.10.

1.5.2.2.2   Multistep Cartesian Braiders

The multistep braiding technique is a unique braiding process that enables Cartesian braiding to fabricate structures with complex shapes such as a T joint, an I beam, and a box beam. In each step, row and column displacements are independently effected along with nonbraider yarn selection on Cartesian braiders with different machine bed configurations and specialized mechanisms. A multistep braiding process has multiple yarn groups and varying numbers of yarns per group. Yarn groups are sets of yarns that follow the same path during the braiding process. A typical eight-step braiding process with yarn grouping and braided architecture is displayed in Figure 1.11 [2].

1.5.2.2.3   Hexagonal Braiders

A novel hexagonal 3D rotary braiding system [27] utilizing hexagonal horn gears, as shown in Figure 1.12, has been developed by Advanced Fibrous Materials

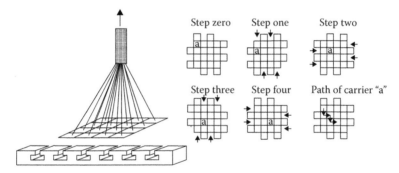

FIGURE 1.10   Four-step Cartesian braiding process. (From Paul, O.I., *Int. J. Eng. Sci.*, 3(3), 01–08, 2014.) [26].

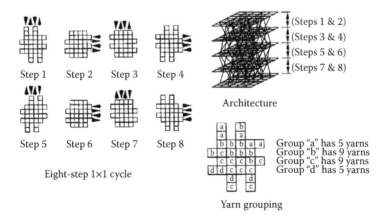

FIGURE 1.11   Multistep Cartesian braiding process. (Kostar, T., and T. Chou, *3-D Textile Reinforcements in Composite Materials*, Woodhead Publishing Ltd. and CRC Press LLC, Cambridge, United Kingdom, 217–240, 1999.) [2]

FIGURE 1.12 Hexagonal 3D rotary braider. (Schreiber et al., Novel three-dimensional braiding approach and its products, ICCM-17: 17th International Conference on Composite Materials, Edinburgh, United Kingdom, July 27–31, 2009, IOM Communications, London, 2009.) [27]

Laboratory (University of British Columbia, Vancouver, Canada) in cooperation with the Institut für Textiltechnik (RWTH Aachen University, Aachen, Germany). This unique cam arrangement allows every cam to carry a maximum number of six carriers placed in 60° intervals around the cam, although only one carrier is allowed to take a mutual position between two adjacent cams, and therefore six different directions of planar carrier movement. The foot part of the carrier is designed as a glider whereby the bobbin glides during movement from one position to the other by using the adjacent carriers as a bearing surface. Compared to the traditional four-bobbin carriers, which are capable of moving carriers in four different planar directions, this unique hexagonal cam arrangement adds two more planar movement directions to the process. It has been observed that the hexagonal 3D braider could produce braided patterns ranging from traditional line/lace or circular braiding to complex 3D braiding patterns like triangle, diamond, or even star shape [27].

1.6 BRAID GEOMETRY

A typical braid geometry is represented in Figure 1.13, which consists of braiding axis, line (L), stitch (S), and braiding angle (θ).

Braid axis is the direction along which the braided fabric is formed. Line can be defined as one repeat of the braid structure perpendicular to the braid axis. The number of lines in a braided structure can be related to number of carriers in the braiding machine. Stitch is defined as one repeat of the structure along the braid axis. Braid angle is the angle formed by the intertwining yarn with the braid axis. Braiding machine speed in terms of movement of carriers and take-up speed determines the braiding angle and number of lines or stitches per unit length.

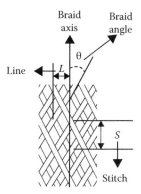

FIGURE 1.13   Braid geometry.

Braiding angle is the most important parameter that determines the cover factor of braided structure. A low carrier speed with a higher take-up speed results in loose braided structures with low braiding angles, which have low cover factor. But high carrier speed and low take-up speed result in closely packed braided structures having higher braiding angles and better cover factors [28].

1.7   APPLICATION OF BRAIDS

Braided fabrics have innumerable applications in textile, civil infrastructure, aerospace, medical, transportation, and electrical fields. The applications in each sector are given here.

The most common textile and garment applications of braids include cords/laces for clothes and shoes. Candlewicks, sash cords, domestic ropes, water ski ropes, mountaineering ropes, yachting ropes, parachute lines, fishing nets, and mooring lines are some of the important technical textile applications. Figure 1.14 illustrates mountaineering ropes having a kernmantle structure [29,30] prepared by using nylon core braided by polyester fiber sheathing providing the rope with excellent strength, low stretch, durability, and flexibility coupled with outstanding abrasion resistance.

Braided reinforcements have found applications in sporting goods like baseball bats, hockey sticks, nets, golf clubs, and so on. Electrical insulations, electrical cables, transportation applications such as truck winch ropes, front end aprons for automobiles, and energy storage flywheels for hybrid electric vehicles are a few of the applications where braided structures are the preferred ones for ropes. Aerospace applications of braided structures include aircraft engine parts and one-piece aircraft horizontal stabilizers. High strength, lightweight, noncorroding, and high stiffness attributes coupled with low manufacturing cost of braided reinforcements have paved the way for many civil infrastructure applications such as trenchless pipes, metallic braided ropes for elevator lifting, bridge ropes, steel braided ropes for cement reinforcement, and inflatable helicopter repair shelters [31].

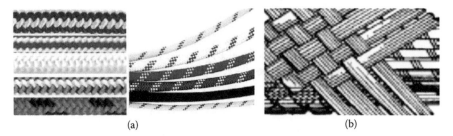

(a)   (b)

FIGURE 1.14 Mountaineering ropes: (a) kermantle braided ropes and (b) kermantle structure. (From Bent et al., Mountaineering ropes of core-mantle structure, U.S. Patent 3,073,209, 1963; Pelican Rope Works, *Static Kernmantle Braided Ropes*, Santa Ana, CA, 2009, retrieved from http://www.pelicanrope.com/staticropes.html, accessed on April 2, 2015.) [29,30]

Medical applications of braids include prosthetic artificial limbs, which have the dual advantage of comfort and conformability [32]. Microbraids using delicate silk fibers have also found applications as stents and other implantable biomaterial applications. Some of the other medical textiles that are manufactured using braiding technology include sutures, biodegradable as well as nonbiodegradable; prostatic stents; braided composite bone plate; bone setting device; overbraided high-pressure tubes; ground cables or harnesses; braided pillar implants; artificial ligament or tendon; artificial cartilage; braided dental floss (catheters); surgical braided cable; braided biomedical tubing; and so on [33,34].

1.8   MERITS AND DEMERITS OF BRAIDING

Important advantages of fabric formation by braiding are cost-effectiveness, excellent strength in all directions, and ability to form complex shapes compared to conventional fabric formation techniques. The development of braiding equipment shows a trend toward automatic fabrication of a variety of braided structures. Issues such as braid convergence, processing cost, time, and braid stability have been overcome. The dominant limiting factors in braiding include the following: the entire supply of braiding yarns (packages or yarn carriers) must be moved, the machine size is large relative to the braidable cross-sectional size of preform, only limited lengths of braid may be formed, the range of fiber architecture is constrained by the process, and different machines are usually required to vary the braiding pattern [35].

The limitations of weaving, knitting, nonwoven, and stitching processes include poor shear resistance, limited strength in the primary loading direction, and inability to produce complex shaped parts, which are easily addressed by the braiding process. Figure 1.15 illustrates the behavior of different fabric structures under uniaxial stress [34]. It portrays good conformability of braided structures to complex shapes, high dimensional stability (in all directions), and in-plane shear resistance along machine ($0°$) and bias directions ($\pm45°$) while low in-plane shear resistance along cross direction ($90°$).

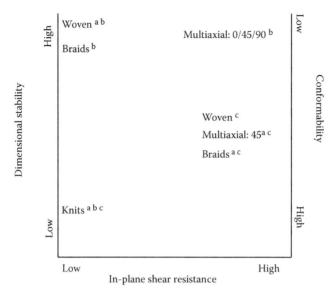

FIGURE 1.15 Behavior of various fabric structural forms under uniaxial stress: (a) machine (0°) direction, (b) cross (90°) direction, and bias (±45°) direction. (From Chou, T.W., and F.K. Ko, *Textile Structural Composites*, Elsevier Science, Amsterdam, Netherlands, 1989.) [35]

1.9 SUMMARY

Braiding process is one of the most cost-effective fabric manufacturing processes. Braided fabrics provide excellent strength for the structure in all directions and have exceptional ability to form complex shapes. Braiding has established itself as the most preferred fabric formation technique for complex 3D geometry creation for applications in aerospace, transportation, civil infrastructure, and medical fields. Recent advances in application of textile fabrics for technical field have witnessed a spurt in 3D braided structure design developments along with improved braiding processes and machineries.

REFERENCES

1. Ko FK (1987). Braiding. *Engineered Materials Handbook*, Volume 1. ASM International, Metals Park, Ohio, 519–528.
2. Kostar T and Chou T (1999). Braided structures. In Miravete, A. (Ed.) *3-D Textile Reinforcements in Composite Materials*. Woodhead Publishing Ltd. and CRC Press LLC, Cambridge, United Kingdom, 217–240.
3. Bicking AM (2011). Explorations in fancy braid creation through the use of industrial machinery (MS Thesis). North Carolina State University, Raleigh, North Carolina, 2011.
4. Owen R (1995). *Braids: 250 Patterns from Japan, Peru, & Beyond*. Interweave Press, Loveland.
5. Good I (2001). Archaeological textiles: a review of current research. *Annual Review of Anthropology*, 30, 209–226.

6. Blakesley K (2010). About simple braided rug making. Retrieved from http://www.ehow.co.uk/about_6579597_simple-braided-rug-making.html.
7. Larsen JL (1986). *Interlacing: the Elemental Fabric*. Kodansha International, Tokyo, Japan.
8. Braley M and Dingeldein M (2001, May). Advancements in braided materials technology. A&P Technology, Inc., 46th International SAMPE Symposium, Long Beach, CA. 2445–2454.
9. Branscomb D, Beale D, and Broughton R (2013). New directions in braiding. *Journal of Engineered Fibers and Fabrics*, 8(2), 11–24.
10. Ueda T, Ohtani A, Nakai A, and Hamada H (2002). Mechanical properties of square braided fabric. *Proceedings of ASME International Mechanical Engineering Congress & Exposition*, Vol., 1–5, 19–23.
11. Tada M (1998). Kumihimo for textile composites. *Proceeding of 4th International Symposium on Textile Composites*, Kyoto, Japan, O-41-1. Retrieved from http://www.kumihimo-society.org/kumihimo/Key14.html.
12. Tada M, Uozumi T, Nakai A, and Hamada H (2001). *Composite Part A*, 32, 1485–1489.
13. Farell, M (2014). *Stainless steel XC expandable braided sleeving*. Techflex, Flexible Technologies Limited, Auckland. Accessed on April 2, 2015.
14. Brunnschweiler D (1953). Braids and braiding. *Journal of the Textile Institute*, 44(1), 666–685.
15. Ma G (2010). Modeling, machine vision sensing and material flow system definition of braiding point motion (Thesis). Auburn University, Alabama.
16. Lundgren J, Braiding Machine. U.S. Patent 887, 257, 1903.
17. Wardwell SW, Braiding Machine. U.S. Patent 1, 032, 870, 1912.
18. Wardwell SW, Braiding Machine. U.S. Patent 1, 197, 692, 1916.
19. Sokol VG, Braiding Machine. U.S. Patent 2, 64, 899, 1949.
20. Haehnel LVA and Rudolf H. Braiding machine. U.S. Patent 4, 765, 220, 1988.
21. Fischer, TA. Speed control apparatus and method for braiding machine, http://www.patentstorm.us/patents/4716807/description.html, U.S. Patent 4716807, issued January 5, 1988.
22. Lombard N. Braiding-machine. U.S. Patent 428,136, August 25, 1890.
23. Douglass W (1964). *Braiding and Braiding Machinery*. Centrex Publishing, Eindhoven, Netherlands. Print.
24. Walter M. Presz, Jr., Stanley Kowalski, III, Wardwell Braiding Machine Company, Powered lower bobbin feed system for deflector type rotary braiding machines, U.S. Patent 7,270,043 B2, 2007.
25. Adanur S (1995). *Wellington Sears Handbook of Industrial Textiles*. Technomic Publishing Company, Inc., Lancaster, Pennsylvania.
26. Paul OI (2014). Modeling and simulation of three dimensionally braided composite and mechanical properties analysis using finite element method (FEM). *International Journal of Engineering and Science*, 3(3), 01–08.
27. Schreiber F, Ko FK, Yang HJ, Amalric E, and Gries T (2009). Novel three-dimensional braiding approach and its products, ICCM-17: 17th International Conference on Composite Materials, Edinburgh, United Kingdom, July 27–31, 2009, IOM Communications, London.
28. Rawal A, Potluri P, and Steele C (2005). Geometrical modeling of the yarn paths in three-dimensional braided structures. *Journal of Industrial Textiles*, 35(115), 115–135.
29. Benk K and Meyer W. Mountaineering ropes of core-mantle structure, U.S. Patent 3, 073, 209, 1963.
30. Pelican Rope Works (2009). *Static Kernmantle Braided Ropes*. Santa Ana, CA. Accessed on April 2, 2015.

31. Thomson DT (1998). Braiding applications for civil infrastructure. In *Special Papers, Session 1: Braiding and Fabrics: Advance Reinforcement Architecture Opens New Horizons*. American Composite Manufacturers Association, Arlington. Retrieved from http://www.acmanet.org/research/spi_papers_90_thru_99/SPI_papers_98/1998-01.pdf.

32. Head AA (1998). Commercial applications for novel braided preform constructions. In *Special Papers, Session 1: Braiding and Fabrics: Advance Reinforcement Architecture Opens New Horizons*. American Composite Manufacturers Association, Arlington. Retrieved from http://www.acmanet.org/research/spi_papers_90_thru_99/SPI_papers_98/1998-01.pdf.

33. Chellamani KP, Sudharsan J, and Sathish J (2013). Medical textiles using braiding technology. *Journal of Academia and Industrial Research (JAIR)*, 2(1), 21–26.

34. Gopalakrishnan D, Bhuvaneswari, Ramakrishnan V, and Sabarinath T (2007). Braiding & narrow width fabrics. *Indian Textile Journal*, 118(1), 97–104.

35. Tsu-Wei Chou and Frank K. Ko (1989). *Textile Structural Composites*, Elsevier Science, Amsterdam.

2 Braiding Process and Parameters

Aron Gabor

CONTENTS

ABSTRACT

This chapter gives an overview of the main characteristics of circular braid manu-facturing processes with special attention to the interaction between process parameters and structural as well as mechanical properties of braided preforms. Circular two-dimensional (2D) and three-dimensional (3D) interlock processes are considered. Analytical modeling approach to understand the influence of pro-cess parameters, corresponding to a transient or a steady-state production phase, on the final structure of braids is presented too. The analysis of the properties of braids is approached using non-destructive tests, such as x-ray tomography, while the mechanical properties are estimated on braided composite tubes using classical testing and extensometry methods. The variation of the mechanical properties as a function of the braiding angle is discussed.

2.1 PRODUCTION OF BRAIDS

Textile processes (weaving, knitting, or braiding) leading to 2D layered fiber structures are largely used for the manufacturing of composite parts for the needs of maritime, automobile, aerospace, or medical engineering. They allow a relatively good production rate, and the products are able to correctly withstand in-plane stress and strain states. Nevertheless, to obtain thicker or complex-shaped preforms it is necessary to stack or to assemble several 2D layers. In this case, alternating relatively stiff and resistant layers with interfaces filled with resin having relatively weak mechanical properties lead to inferior mechanical behavior (Figure 2.1). Damaging occurs due to delamination or cohesive behavior, mainly under fatigue, shock, or shear loading, and the resin acts as the weakest link. Consequently, academic and industrial sectors are continuously developing advanced textile processes for the manufacturing of 3D textile architectures to achieve high mechanical performance and high resistance.

As stated in the works of Stover et al. (1971) and Mouritz et al. (1999), braiding technology was used for the very first time in aerospace industry to produce preform for a composite structure (Figure 2.2). A four-step Cartesian braiding process was developed to produce 3D carbon–carbon motor components to achieve substantial weight savings.

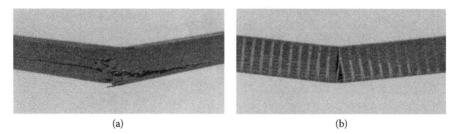

(a)   (b)

FIGURE 2.1 Bending behavior of composites. (a) Laminated multilayer composite. (b) Composite produced using 3D manufacturing process.

(a)   (b)   (c)

FIGURE 2.2 Composite architectures obtained by different manufacturing techniques. (a) Weaved shield. (b) 2D Braided rocket nozzle. (c) Braided connectors. (From Mouritz, A. et al., *Composites Part A*, 30(12), 1445–1461, 1999.)

A Cartesian braiding machine is made of a plane platform, where spindles/spools translate in a well-organized manner, to form a braid. The basic Cartesian process involves four distinct Cartesian motions of groups of yarns called rows and columns. The spindles are generally moved by pneumatic jacks. For a given step, alternate rows (or columns) are shifted to a defined distance relative to each other (Figure 2.3). The next step involves the alternate shifting sequence of the first and second steps, respectively. A complete set of four steps is called a machine cycle. It should be noted that after one machine cycle the rows and columns are returned to their original positions. The interlocking of yarns occurs at half cycle. Cartesian braiding process allows to obtain through-the-thickness braided composites of different shapes such as I, L, and so on (Tada et al., 2001). The control parameters of the process are: (1) the x and y shift step of the spindles, (2) the braiding angle, and (3) the nature and type of the fiber. A complete description of this braiding process is given by Ko (1987) and Brown and Crow Jr. (1992). Different types of braiding processes are also discussed in detail in Chapter 1 of the present book.

Another type of braiding that has been developed is the circular braiding process (Lee, 1993; Miravete, 1999). This process has the ability to produce hollow and circular shapes in a single step to meet the requirements of several industrial applications. Major limitations in the design of machine-made braids include restricted width, diameter, thickness, and shape selection. In the case of special or batch braids, the main limitations are productivity and product length.

In a basic circular braiding process, the aim is to place yarns in right- and left-handed interlaced helices. The action and mechanisms of the braiding machine should perform the interlacing. Therefore, the main component of a circular braiding machine is a circular platform on which carriers are placed to hold the spools of yarn as shown in Figure 2.4. The number of spools/carriers depends on the desired density of the

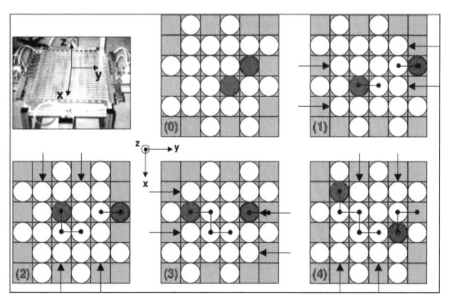

FIGURE 2.3   Steps of a Cartesian braiding process.

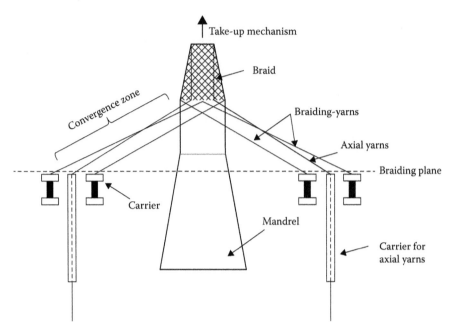

FIGURE 2.4 Scheme of a circular braiding process.

braid. Additionally, axial or core yarns can be added to this structure as longitudinal reinforcement or usually as a solid, removable component (the mandrel) to achieve the final shape of the braid. The yarns are held together instantaneously in the braid formation point, generally on the surface of the mandrel, forming a self-supporting structure. From the kinematic point of view, it can be considered that one end of a yarn moves freely on the platform, whereas the other one is held in place on the mandrel. The two main control parameters of the production process are (1) the rotational speed of the carriers on the platform, and (2) the translational velocity of the mandrel. When special-shaped mandrels are used, other parameters, such as eccentricity or rotation of the mandrel, can also have an influence on the architecture of the final braid.

There are two types of circular braiding process, 2D braiding and 3D multilayer interlock braiding. From the manufacturing process point of view, the main difference between these two processes is the motion path of the spindles on the braiding platform: In the case of 2D braiding, the yarns are interlaced in their plane, while during 3D braiding, several layers are continuously interlaced through the thickness.

Both processes offer the ability to manufacture a range of braided architectures that can be adapted to specific structural applications by providing the optimum amounts of axial and through-thickness yarns (Figure 2.5).

2.1.1 2D CIRCULAR BRAIDING

In the case of 2D braiding, the braiding platform contains only one level (crown) of carriers, and the carriers move in a circular and sinusoidal way relative to the platform due to the action and mechanisms of a horn-gear system (Figure 2.6a and b).

FIGURE 2.5   Two-dimensional circular braiding using a square-shaped mandrel.

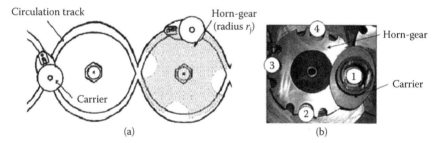

FIGURE 2.6   (a) Motion of carriers with the action of horn-gears. (b) Possible positions of a carrier on a horn-gear.

The carriers are passed from one position to another and lead the yarns to form a crossed pattern in the plane of the braid. This type of braiding can be considered as a circular weaving, where a closed-waved textile is obtained.

The warp carriers move in the counterclockwise direction, and the weft carriers move in the clockwise direction. Usually, the mandrel is placed in the center of the braiding platform normal to its plane. The translational movement of the mandrel allows the braid to be constructed on the entire surface of the mandrel. The circular motion of the carriers around the braiding axis rolls up the yarns on the mandrel, while the sinusoidal movement of the carriers leads to the interlacing of the yarns. The yarns leaving the spools interlace progressively, forming a convergence zone between the plane where the yarns leave the carriers and the braiding front situated at the contact point with the mandrel.

2.1.2   3D Circular Braiding

In the case of 3D interlock circular braiding, there can be several coaxial crowns of carriers (Figure 2.7a). It is possible with a switching system to pass the carrier from one crown to another; therefore, the carrier moves along the height of the

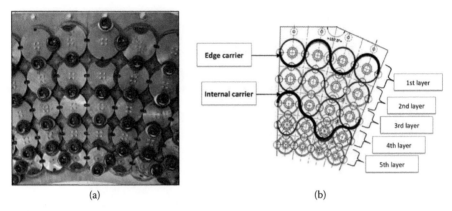

FIGURE 2.7 (a) Overview of a multilevel horn-gear and carrier system for 3D braiding. (b) Motion of carriers during a 3D braiding process. In this case, five layers are interlocked two by two.

braiding platform. Some carriers moving by two levels produce an interlocking of layers (Figure 2.7b). Other internal and external carriers move by one level and as a consequence, close the braiding path on the platform. The interlock braiding allows the introduction of core or axial yarns, coming from the center of the horn-gears, without any spinning motion.

2.2 MECHANICS OF BRAIDING PROCESS

The braiding process involves different mechanisms that influence the positioning of the yarns on the mandrel (Du and Popper, 1994; Kessels and Akkerman, 2002; Michaeli et al., 1990; Zhang et al., 1999a) (Figure 2.8). If process parameters are maintained constant for a sufficient period, equilibrium occurs between the translation speed of the mandrel and the rolling-up speed of the yarns on the mandrel. The lay-down spot of the strands, the braiding front, occurs on the same relative position measured from the braiding platform. Until this equilibrium exists, the braiding process remains in a steady state.

Therefore, the invariability of the position of the braiding front may be used to describe the steady-state process. When the process parameters are changed, this equilibrium is broken and only a new value of the braiding angle will allow to restore the equilibrium between the translation speed of the mandrel and the rolling-up speed of yarns. The difference between the first and second braiding angle will vary the position of the braiding front. Until this variation exists, the braiding process can be considered as transitory.

The system naturally tends to return to its steady-state phase. If the new process parameters are maintained constant for a sufficiently long time, the braiding angle vary progressively from its actual value to a new one, proper to the new steady-state run. Accordingly, the motion speed of the braiding front tends to a zero value.

After leaving the carriers, the yarns are subjected to frictional interactions in the convergence zone during the interlacing. The friction between the yarns modifies

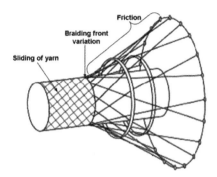

FIGURE 2.8   Main phenomena occurring during braiding process.

their trajectory and reduces the rotational speed of the yarns around the braiding axis and consequently influences the final parameters of the braid (e.g., the braiding angle). Nevertheless, a rectilinear path hypothesis is largely adopted when modeling the process.

During the braiding process, the yarns are stretched, and the tangential component of the tensile force can lead to the sliding of yarns on the mandrel. In the same way, the rolling-up of yarns on the mandrel leads to a normal force (pressure) on the surface of the mandrel, opposing the sliding of yarns. The equilibrium between tangential and normal forces can be used as a condition to define the sliding limit. In the case of particular mandrel shapes, it is possible that the normal force gets oriented outside, leading to the detachment of the yarns from the surface of the mandrel.

2.3   PROCESS PARAMETERS

First, in the hypothesis that the trajectory of the yarns in the convergence zone is rectilinear, the yarns will follow the motion of the carriers on the braiding platform. It is assumed that the carriers have an overall rotational motion around the braiding axis describing the circumference of the braiding platform. Two process parameters have to be considered.

1. The overall rotational speed Ω, which is linked to the spinning speed ω of the horn-gears and their number N by the equation (Guyader, 2012)

$$\Omega = \frac{\omega}{2N} \tag{2.1}$$

The radius of this overall rotational movement is denoted by R_g. When a braiding ring is used, the yarn motion is defined accordingly, and the dimensions of the braiding platform only are substituted by those of the braiding ring.

2. The translational speed of the mandrel, denoted by V. The mandrel motion directly affects the formation of the braid; generally, only a translational motion, normal to the braiding platform, is applied to the mandrel. However, a

translational motion is not sufficient for the braiding of complex-shaped man-drels. To avoid the eccentricity of the mandrel during braiding operation, the center of the mandrel cross section at the braiding front is maintained on the braiding axis. This means that, for mandrel geometries without rectilinear axis, the motion applied to the mandrel has to be complex, not only a translational one. Often it is necessary to set up an automated process using, for example, robotized pneumatic arms.

2.4 GEOMETRICAL PARAMETERS

At mesoscopic level, the geometric characteristics of the braid can be given by the orientation and lay down of constitutive yarns. In case of a 2D braid, three param-eters are involved (Figure 2.9), as discussed below:

- The first parameter, known as the braiding angle and noted by $\dfrac{\alpha}{2}$, gives the orientation of the yarns in the plane of the fabric. It can be defined as the angle between the yarn and the projection of the braiding axis on the sur-face of the mandrel.

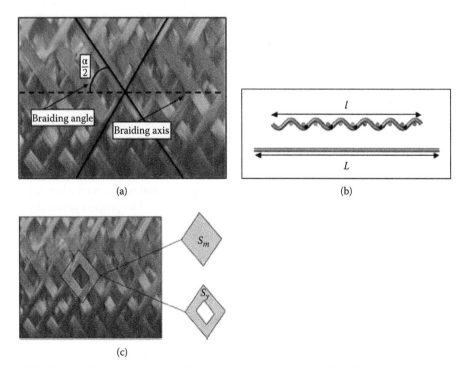

(a)

(b)

(c)

FIGURE 2.9 Definition of geometrical parameters of a braid: (a) Braiding angle, (b) shrink-age, and (c) covering factor.

- The second parameter is the shrinkage, denoted by e, which characterizes the weaving of the yarn over the thickness of the fabric. It is calculated from the length of the fabric l and the initial length L of the yarn as follows:

$$e = \frac{L-l}{L} \times 100 \tag{2.2}$$

- The third parameter is the coverage factor, denoted as C, calculated from the ratio between the area given by the yarns S_y and the total area of the surface of the mandrel S_m as follows:

$$C = 1 - \frac{S_y}{S_m} \tag{2.3}$$

The coverage factor is an adapted indicator for the permeability of the preform.

2.5   INFLUENCE OF PARAMETERS ON STRUCTURE

One of the most important parameters of the geometry of a braid is the braiding angle, directly related to the mechanical properties of the obtained composites. For the control of braiding process, it is important to establish relationships between the process parameters and the braiding angle. As stated earlier in this chapter (Section 2.3), the two main process parameters are the overall rotational speed of the braiding platform, denoted by Ω, and the translational speed of the mandrel, denoted by V. These parameters can be related to the braiding angle α and the complexity of these relationships depends on the adopted hypotheses, which describe the different phenomena occurring during braiding. Generally, three main hypotheses are considered in the literature.

- A rectilinear path of the yarns in the zone located between the carrier platform and the mandrel, neglecting the friction between the yarns during the interlacing
- A perfect grip of the yarns on the mandrel, considering no relative movement of yarns on the mandrel
- A circular trajectory of the carriers on the braiding platform, neglecting their sinusoidal movement around an average circle

2.5.1   RELATIONSHIPS BETWEEN THE PROCESS PARAMETERS AND THE GEOMETRY OF THE BRAID

To obtain a better approach to understand the structure of the braid, Kessels (Kessels and Akkerman, 2002) proposed to model the sliding of the yarns on the mandrel, Zhang (Zhang et al., 1999a,b) modeled the friction between the yarns in the convergence zone, considering a modified yarn trajectory, and Du (Du and Popper, 1994) introduced the transitory movement of the braiding front in the case where the braiding parameters are modified during the process.

At the beginning of these studies, Ko (1987) proposed an analytical equation for the braiding angle in case of a symmetric circular mandrel.

$$\frac{\alpha}{2} = \tan^{-1}\left(\frac{R\Omega}{V}\right) \tag{2.4}$$

where R is the radius of the mandrel, Ω is the rotational speed of the braiding platform, and V is the translational speed of the mandrel.

Similarly, in the work of Du and Popper (1994) an analytical model is developed to describe transitory state runs. If process parameters change, this will affect the equilibrium between the rolling-up speed and translation speed of the mandrel. This disequilibrium can be observed as a variation of the position of the braiding front. Du divides the surface of the mandrel in cone-shaped elements. Considering that the braiding angle α is equal to the angle formed by the yarn and the elementary cone, and projecting the equation of the yarn on the plane of the braiding platform he obtained a relationship for the braiding angle as a function of the position of the braiding front.

$$\tan \alpha = \frac{R_g}{h(t)} \cos \gamma_z \sqrt{1 - \left(\frac{R}{R_g} + \frac{h(t)}{R_g} \tan \gamma_z\right)^2} \tag{2.5}$$

where R is the radius of the mandrel, R_g is the radius of the braiding platform, h is the position of the braiding front and γ_z is the orientation angle of the elementary cone. As the position of the braiding front is time dependent and known at $t = 0$, it is necessary to proceed by an incremental approach, obtaining progressively the value of the braiding angle on the entire surface of the mandrel.

In the case of complex-shaped mandrels, Michaeli et al. (1990) and Akkermann and Kessels (Kessels and Akkerman, 2002) use advanced vectorial models to describe the positioning of the yarns on the mandrel; the braid is modeled as a set of points. Their positions are defined by vectors. The implemented incremental approach is computer assisted.

To establish the influence of process parameters on the structure of the braid in a general case, it is possible to combine the above analytical and vectorial models developed by Du and Michaeli (Guyader, 2012; Guyader et al., 2013). Describing the surface of the mandrel by the radii of curvature and the direction angles of an elementary surface and defining the trajectory of the yarns by a parametric curve, it is possible to establish the contact point of a yarn on the surface of the mandrel as a function of time.

It is necessary to consider several coordinate systems: a Frenet system $(\vec{e}_n, \vec{e}_t, \vec{e}_k)$ attached to the parametric curve and two cylindrical coordinate systems, one attached to the braiding platform $(\vec{e}_r, \vec{e}_\theta, \vec{e}_z)$ and the second one related to the carrier on the braiding platform $(\vec{e}_R, \vec{e}_\Omega, \vec{e}_z)$ (Figure 2.10a).

An elementary surface of the mandrel is described by the distance R between the elementary surface and the braiding axis and the projections of the radii of curvature ρ_z and ρ_θ onto the $(\vec{e}_z, \vec{e}_r)$ and $(\vec{e}_\theta, \vec{e}_r)$ planes, respectively (Figure 2.10b). These radii define the centers of curvature of the elementary area P_z and P_θ, respectively. If we

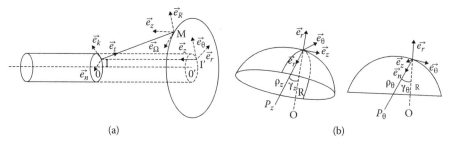

FIGURE 2.10   (a) Coordinate systems used for the description of the geometrical and process parameters. (b) Parameters of an elementary surface of the mandrel.

project the normal axis $\vec{e_n}$ of the elementary area in the plane $(\vec{e_z},\vec{e_r})$, we obtain a longitudinal orientation axis $\overrightarrow{e_{n,z}}$.

Therefore, the orientation of the elementary area in the longitudinal direction can be defined by the angle γ_z formed by the axes $\overrightarrow{e_{n,z}}$ and $\vec{e_r}$. Similarly, we can define the orientation angle γ_θ by projecting the axis $\vec{e_n}$ on the radial plane formed by the axes $(\vec{e_\theta},\vec{e_r})$.

On the defined elementary surface it is possible to divide the yarn trajectory into two components, a longitudinal component $l_z(t)$ and a radial one $l_\theta(t)$. We obtain the analytical equation of the braiding angle as a function of the longitudinal and radial components of the yarn trajectory:

$$\frac{\alpha}{2} = \tan^{-1}\left(\frac{l_\theta(t)}{l_z(t)}\right) \tag{2.6}$$

The radial component can be associated with the rolling-up of the yarn on the mandrel, whereas the longitudinal component corresponds to the translation of the yarns on the mandrel due to its motion and that of the braiding front. To obtain a comprehensive view of the process parameters, a relationship is established between the radial component of the yarn trajectory $l_\theta(t)$ and the translation speed h of the braiding front in the longitudinal direction.

With the assumption of non-slippage of yarns on the mandrel, the path of the yarn is given by the movement of the intersection point $I(t)$ between the yarn trajectory and the surface of the mandrel. We define the points $O(t)$ and $O'(t)$ as the intersection of the braiding axis with the braiding front and with the braiding platform, respectively. The point $I'(t)$ is the projection of the intersection point $I(t)$ on the braiding platform, and the point $M(t)$ represents a carrier (Figure 2.11a).

The radius of curvature ρ_s of the parametric curve at the point $I(t)$ can be given using Euler's theorem:

$$\frac{1}{\rho_s} = \frac{\cos^2\left(\dfrac{\alpha}{2}\right)}{\rho_z} + \frac{\sin^2\left(\dfrac{\alpha}{2}\right)}{\rho_\theta} = \frac{\rho_z\rho_\theta}{\rho_z\sin^2\left(\dfrac{\alpha}{2}\right)+\rho_\theta\cos^2\left(\dfrac{\alpha}{2}\right)} \tag{2.7}$$

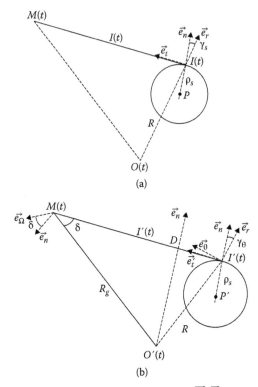

FIGURE 2.11 (a) Projection on the plane defined by axes $(\vec{e}_n, \vec{e}_t)$. (b) Projection of the same elements on the plane of the braiding platform $(\vec{e}_R, \vec{e}_\Omega)$.

The radius of curvature of the parametric curve defines the center of curvature P around which the yarn is rolled up. The distance $l(t)$ represents the length of the yarn between the intersection point $I(t)$ and the carrier $M(t)$. The position of the carrier on the braiding platform $O'M(t)$ is given via the intersection point $I(t)$.

$$\overrightarrow{O'M(t)} = \overrightarrow{O'O(t)} + \overrightarrow{OM(t)} \tag{2.8}$$

with

$$\overrightarrow{OM(t)} = \overrightarrow{OP} + \overrightarrow{PI(t)} + \overrightarrow{IM(t)}$$

$$\overrightarrow{OM(t)} = \overrightarrow{OP} - \rho_s\vec{e}_n + l(t)\vec{e}_t$$

As the yarns roll up on the mandrel, we can assume that the center of curvature P is an instantaneous center of rotation of the yarn. Therefore, it assumes a rotation of the Frenet reference system around its axis $\vec{e}_k$ with a rotational speed denoted by $\dot{\varphi}$. The variation of the braiding angle can be considered as a rotation of the

Frenet reference system around its normal axis $\vec{e_n}$ with a rotational speed denoted by $\left(\dfrac{\alpha}{2}\right)$. With these assumptions, and considering Equation 2.8, a new relationship is obtained connecting the process parameters and the rolling-up speed of yarns on the surface of the mandrel:

$$\Omega R_g \vec{e_\Omega} = V\vec{e_z} + (\dot{l}(t) - \rho_s \dot{\phi})\vec{e_t} + l(t)\dot{\phi}\vec{e_n} + l(t)\left(\frac{\alpha}{2}\right)\vec{e_k} \tag{2.9}$$

The above relationship involves terms expressed in the Frenet reference system and also terms expressed in the reference system of the carrier. To overcome this inconvenience, an elementary arc of the cross section of the braid is projected along a yarn on a plane coinciding with the braiding platform, as shown in Figure 2.11b.

In this figure, the point P', the distance $l'(t)$, and the axis $\vec{e'(t)}$ are defined as the projections on the braiding platform, following the axis $\vec{e'z}$ of the point P, the distance $l(t)$, and the tangential axis $\vec{e_t}$, respectively. The angle δ is the angle formed between the $\vec{e't}$ and $\vec{e_R}$ axes. The equation for the angle δ and the distance $l'(t)$ are easily obtained from the triangle $O'MD$ and can be written as

$$\cos\delta = \frac{\sqrt{R_g^2 - (R + h\tan\gamma_z)^2 \cos^2\gamma_\theta}}{R_g} \tag{2.10}$$

$$l'(t) = R_g\cos\delta - (R + h\tan\gamma_z)\sin\gamma_\theta \tag{2.11}$$

The axes $\vec{e_z}$ and $\vec{e_\Omega}$ of the carrier reference system in the Frenet reference system are written as follows:

$$\vec{e_z} = \cos\gamma_z \vec{e'_z} + \sin\gamma_z \vec{e_n} \tag{2.12}$$

with

$$\vec{e'_z} = \cos\left(\frac{\alpha}{2}\right)\vec{e_t} + \sin\frac{\alpha}{2}\vec{e_k}$$

$$\vec{e_\Omega} = \cos\delta\vec{e_n} + \sin\delta\vec{e'_t} \tag{2.13}$$

with

$$\vec{e'_t} = \cos\left(\frac{\alpha}{2}\right)\vec{e_k} + \sin\left(\frac{\alpha}{2}\right)\vec{e_t}$$

By substituting the relationships of the axes $\vec{e_z}$ and $\vec{e_\Omega}$ into the general Equation 2.9 and considering their projections on the axes of the Frenet reference system, three independent relationships are obtained.

The first, projected on the normal axis $\vec{e}_n$, gives the equation for the rolling-up speed φ of the yarn on the surface of the mandrel:

$$\dot{\varphi} = \frac{\Omega R_g \cos\delta - V \sin\gamma_z}{l(t)} \tag{2.14}$$

with

$$l(t) = \frac{R_g \cos\delta - (R + h\tan\gamma_z)\sin\gamma_\theta}{\sin\left(\dfrac{\alpha}{2}\right)} \tag{2.15}$$

and

$$\cos\delta = \frac{\sqrt{R_g^2 - (R + h\tan\gamma_z)^2 \cos^2\gamma_\theta}}{R_g} \tag{2.16}$$

The rolling-up speed is the ratio between the rate of the deposited length of yarn and the radius of the mandrel.

The second relationship, projected on the tangential axis $\vec{e}_t$, connects the variation of the length of yarn $\dot{l}(t)$ between the carrier and the intersection point to the process parameters and the braiding angle:

$$\dot{l}(t) = \Omega R_g \sin\delta \sin\left(\frac{\alpha}{2}\right) - V\cos\gamma_z \cos\left(\frac{\alpha}{2}\right) + \rho_s\varphi \tag{2.17}$$

The third equation projected on the transversal axis $\vec{e}_k$, gives the braiding angle variation as a function of the process parameters and the actual braiding angle:

$$\left(\frac{\dot{\alpha}}{2}\right) = \frac{\Omega R_g \sin\delta \cos\left(\dfrac{\alpha}{2}\right) - V\cos\gamma_z \sin\left(\dfrac{\alpha}{2}\right)}{l(t)} \tag{2.18}$$

Equations 2.14, 2.17, and 2.18 are able to describe several phenomena, such as the slippage of the yarn on the mandrel or the relaxation of the yarn during the braiding operation.

The first equation leads to the equations for the radial component of the yarn trajectory and to the braiding front translation speed $\dot{h}$. Assuming the center of curvature P as an instantaneous center of rotation, the motion of the intersection point $I(t)$ can be considered as a rotation around the center of curvature of the tangential axis with a rotational speed $\dot{\varphi}(t)$. The tangential component of the deposition of yarn on the mandrel $l_t(t)$ is consequently expressed as

$$l_t(t) = \rho_s\dot{\varphi}(t) \tag{2.19}$$

This equation is divided into radial and longitudinal components that can be easily expressed by projection on the radial and longitudinal axes, respectively.

$$l_\theta(t) = l_t(t)\sin\frac{\alpha}{2}$$

$$l_e(t) = l_t(t)\cos\frac{\alpha}{2}$$

Hence, the radial component of the yarn trajectory $l_\theta t$ and the braiding front translation speed h in the longitudinal direction are obtained as

$$l_\theta(t) = \rho_s\Omega\frac{R_g\cos\delta - V\sin\gamma_z}{l(t)}\sin\frac{\alpha}{2} \tag{2.20}$$

$$\dot{h} = V - \rho_s\Omega\frac{R_g\cos\delta - V\sin\gamma_z}{l(t)}\cos\frac{\alpha}{2} \tag{2.21}$$

Assuming that the radial component of the yarn trajectory on the mandrel corresponds to the rolling-up speed of the yarn and the longitudinal component corresponds to the translation of the yarn due to the motion of the mandrel and the braiding front, we can construct a general relationship of the braiding angle adapted to complex mandrel shapes and transitory process run.

$$\frac{\alpha}{2} = \tan^{-1}\left(\frac{\rho_s\dot{\phi}(t)}{V+\dot{h}}\right) \tag{2.22}$$

Equation 2.22, in case of a circular mandrel of radius R and a steady-state run, $h = 0, \gamma_\theta = 0, \gamma_z = 0, \cos\delta = \dfrac{\sqrt{R_g^2 - R^2\cos^2\gamma_\theta}}{R_g} = 1, \rho_s = \dfrac{R}{\sin^2\dfrac{\alpha}{2}}$, gives Equation 2.4 of Ko:

$$\frac{\alpha}{2} = \tan^{-1}\left(\frac{R\Omega}{V}\right)$$

Considering the same mandrel but in a transitory-state run, the following relations are obtained:

$$l_\theta(t) = R\Omega \tag{2.23}$$

$$\dot{h} = V - \frac{R\Omega h(t)}{\sqrt{R_g^2 - R^2}} \tag{2.24}$$

which lead to the braiding angle relationship given by Zhang et al. (1999a).

$$\frac{\alpha}{2} = \tan^{-1}\left(\frac{\sqrt{R_g^2 - R^2}}{h(t)}\right) \tag{2.25}$$

On the other hand, expressing the speed of the braiding front in Equation 2.22, Equation 2.5 given by Du and Popper is obtained:

$$\dot{h} = V - \frac{R\Omega h(t)}{R_g\sqrt{\frac{1}{R_g} - \left(\frac{R}{R_g} + \frac{h(t)}{R_g}\tan\gamma_z\right)^2}} \tag{2.26}$$

In the case of steady-state run with the assumption of tangential deposition of yarns on the mandrel, Zhang et al. (1999b) proposes the following equation of the position of the braiding front $h(t)$:

$$h(t) = \sqrt{R_g^2 - R^2}\,\frac{V}{\Omega R} \tag{2.27}$$

which when substituted in Equation 2.26 of the braiding front motion gives $\dot{h} = 0$.

2.6   FROM A 2D TO A 3D INTERLOCK BRAIDING PROCESS

The previous equations are developed for 2D circular braiding operations, assuming a circular trajectory of the carriers on the braiding platform and neglecting their sinusoidal movement around an average circle. Considering the orientation of the yarn and the thickness of the fabric, an analogy can be constructed between a 3D circular braid and a stack of 2D circular braids (neglecting the interlocking between the superposed layers). In reality, for a 3D circular braiding machine, the transverse motion of the carriers around an average circle may be important, and the assumption of circular trajectory of the carriers on the braiding platform is not obvious. It is noteworthy that the average rotational speed of the carriers is similar to that of the 2D braiding process, and the motion of the carriers from one level to another is quick enough to be averaged by the overall circular motion of the braiding front. Consequently, it can be assumed that the average longitudinal component of the yarn trajectory is not affected.

To validate this hypothesis, it is necessary to verify that the local motion of the carriers has no influence on the motion of the braiding front in a general, transitory-state run.

To establish a transitory process run during the braiding operations, the translational speed of the mandrel is varied and the rotational speed of the braiding platform is held constant. Thus, the position of the braiding front can be varied from an initial value to a set-point value. Both values can be calculated using Equation 2.27, corresponding to the position of the braiding front in a steady-state process run.

Furthermore, it can be stated that in the case of a constant rotational speed of the platform, a mesh develops in the braid during a constant time interval, independent of the other process parameters. For example, in case of a braid having an architecture similar to a taffeta fabric, a complete revolution is needed to create a mesh. Therefore, for a given rotational speed Ω of the platform, a mesh is constructed during a time interval Δt:

$$\Delta t = \frac{4\pi}{\Omega} \tag{2.28}$$

Consequently, by comparing the distance between a defined number of meshes (Figure 2.12a) and the distance traveled by the mandrel, it is possible to establish the motion speed of the braiding front:

$$\dot{h} = V - \frac{d_n}{n\Delta t} \tag{2.29}$$

where $\dot{h}$ is the motion speed of the braiding front, V is the speed of the mandrel, and n is the number of successive meshes.

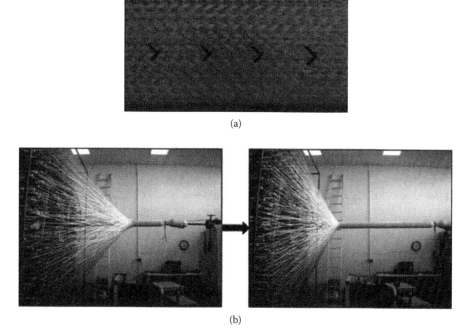

(a)

(b)

FIGURE 2.12 (a) Spacing of meshes during a transitory process run. (b) Motion of the braiding front during a transitory process.

2.6.1   Measurement of the Motion of Braiding Front

2.6.1.1   2D Circular Braiding

To assess the variation of the braiding front during a transitory process, braiding operations were set up using 64 carriers, an effective radius of the braiding platform of 750 mm, a rotational speed of the platform of 140.6 rpm, and a polyvinyl chloride (PVC) mandrel having length of 1000 mm and a diameter of 100 mm (Guyader et al., 2013). The mandrel was moved by a six-degree-of-freedom pulling device driven by a personal computer unit (Figure 2.13a). The mandrel was translated normal to the braiding front. The same type of fiber was used for all braiding operations: E-glass Hybon 2001 Roving and PPG 2400 tex. Filaments had a diameter of 25 µm.

Before starting the measurement, the braiding front was positioned manually on the mandrel at its theoretical value using a laser level; the braiding was started at

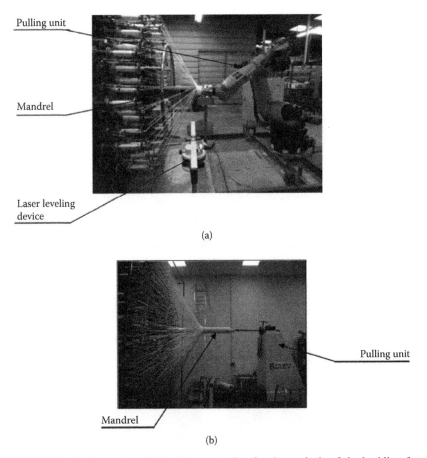

Pulling unit

Mandrel

Laser leveling device

(a)

Pulling unit

Mandrel

(b)

FIGURE 2.13 (a) Setup of a 2D braiding operation for the analysis of the braiding front motion. (b) Setup of a 3D braiding operation.

a constant velocity to achieve stabilization of the braiding front. Next, the motion speed of the mandrel was shifted directly to the set-up value, resulting in the variation of the position of braiding front (Figure 2.12b). As shown in the figure, the contact point between yarns and the mandrel (the braiding front) gets closer to the crown of the braiding machine, and consequently the braiding angle changes.

2.6.1.2 3D Circular Braiding

To analyze the motion of the braiding front during a transitory-state process, a 3D braiding platform was set up, consisting of five carrier levels that define an effective internal and external diameter of 2000 and 3000 mm, respectively. Each carrier level was composed of 64 carriers, the rotational speed of the braiding platform was 140 rpm, and the mandrel was made of a PVC pipe with length of 2000 mm and diameter of 50 mm (Figure 2.13b). An automated pulling unit allowed the control of speed variation and the mandrel movement normal to the plane of the braiding platform. Figure 2.14 presents the experimental and theoretical results for two 2D braiding operations, the first one when the speed of the mandrel is increased and the second one when mandrel speed is decreased. In Figure 2.15, two 3D braiding situations are presented in which the mandrel motion's speed was changed from 13.12 mm/s to 18.75 mm/s and from 8.44 mm/s to 6.19 mm/s.

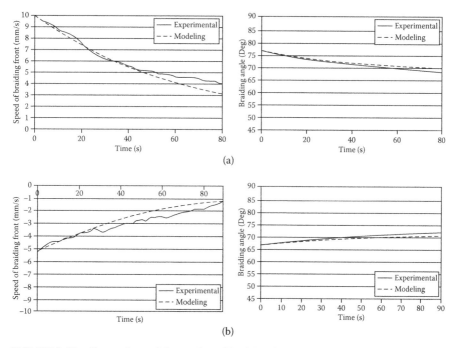

FIGURE 2.14 Comparison of the motion of braiding front (left) and of the braiding angle (right) for the braiding processes carried out by (a) accelerating the translation speed of the mandrel from an initial value of 5 mm/s to 15 mm/s and (b) by slowing down the translation speed of the mandrel from an initial value of 10 mm/s to 5 mm/s.

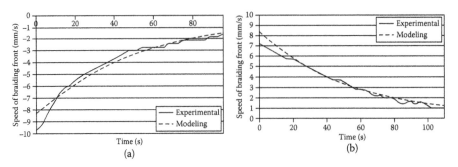

FIGURE 2.15 Comparison of the translation of braiding front in case of 3D circular braiding when the mandrel speed (a) decreases and (b) increases.

2.7 ANALYSIS OF THE STRUCTURE OF BRAIDS

Depending on the impregnation technique (resin transfer molding [RTM], injection, or compression molding), the architecture of the braid can be modified. Nondestructive techniques can be used successfully to get an overview of the structure of the braid (Guyader, 2012). For example, tomography allows studying braided coupon specimens, the dimensions of which correspond to a representative volume element of a mesh. Tomography by x-ray absorption can gather projection data from multiple directions and feeds the data into a tomographic reconstruction software algorithm processed by a computer (Figure 2.16a). With the acquired data, it is possible to construct a 3D digital image, transposing the diminution of x-ray power in gray shades. The density level of the gray shades allows to get an overview of the braid. The sequence of the obtained 2D images can be used to control the distribution of yarns over the height of the braid. Using digital image processing, a color code can be attributed to the different yarns, as a function of their orientation. Figure 2.16b compares the braiding angle over the height of a composite for two braiding angles, 35 degrees and 65 degrees. The cost of tomography tests could be quite important, and therefore it is recommended to do this type of test at the launch of a new process or a new product, to adapt the process parameters to the requirements of the final product.

2.8 TESTING METHODS FOR MECHANICAL PROPERTIES

Generally, two aspects of the mechanical behavior of braided composite materials are investigated: mechanical parameters related to stiffness (e.g., elastic and shear moduli) and failure behavior (e.g., strength and damage models). These parameters can be determined under a wide range of tests, such as tension, compression, torsion, or internal/external pressure.

For tensile test purposes, flat coupon specimens are the most widely used specimen type, and the obtained mechanical properties are considered as the representative of the final composite products (Cox et al., 1994; Falzon and Herszberg, 1998; Kohlman et al., 2012). This approach can be criticized because it neglects the differences between manufacturing process of flat and tubular composites. It is difficult

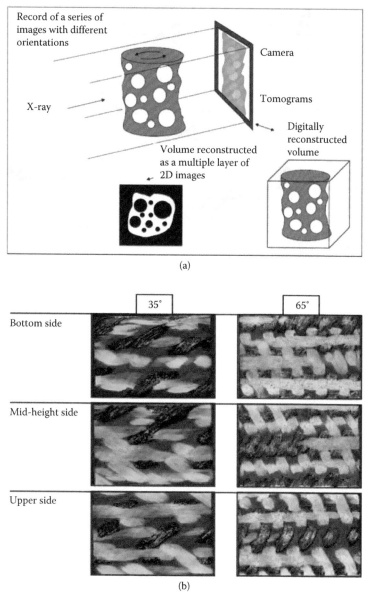

(a)

(b)

FIGURE 2.16 (a) Principles of x-ray tomography. (b) Digitally processed tomograms of two types of glass fiber braids.

to obtain tubular specimens with constant thickness, and testing a cut out coupon can lead to different results from the real behavior. Also, the continuity of fibers is broken by the cutting operation, leading to undesirable edge effects.

In case of low-diameter braided tubes (10–40 mm), it is possible to test directly the braid in tension, bending, or torsion to obtain stiffness parameters and to also study ultimate behavior (de Oliveira Simões and Marques, 2001; Potluri et al., 2006;

Zsigmond, 2005). Compression tests are used to evaluate crushing behavior or inter-laminar behavior (Calme et al., 2006; Chiu, 1999).

Depending on the application domain (e.g., vehicle or air transportation industry), complementary low- or high-velocity impact tests are performed to establish consti-tutive equations and energy absorption properties (Hamada et al., 1999; Priem et al., 2014; Subhash et al., 2001; Sutcliffe et al., 2012).

To measure strain field, classical extensometry is generally used, but the develop-ment of digital image processing allows to use new techniques, such as digital image correlation (DIC) (Leung et al., 2013; Pickett and Fouinneteau, 2006).

The loading of a specimen during an elementary test (tensile, compression, etc.) allows establishing a relationship between the stress and strain fields via constitutive equations. From the knowledge of loading conditions and the response of extensomet-ric devices (strain gauges, load cells, displacement transducers, etc.), the parameters of the constitutive equations of the material can be given. When stiffness is analyzed, the response of the composite material is considered generally as linear elastic until the failure. Thus, generalized Hooke's law can be applied to establish the elastic constants, which, in case of an orthotropic composite submitted to a plane stress state, can be written as

$$
\begin{pmatrix} \varepsilon_L \\ \varepsilon_T \\ \sqrt{2}\gamma_{LT} \end{pmatrix} = \begin{bmatrix} \dfrac{1}{E_L} & -\dfrac{v_{LT}}{E_L} & 0 \\ -\dfrac{v_{LT}}{E_L} & \dfrac{1}{E_T} & 0 \\ 0 & 0 & \dfrac{1}{2G_{LT}} \end{bmatrix} \begin{pmatrix} \sigma_L \\ \sigma_T \\ \sqrt{2}\sigma_{LT} \end{pmatrix} \tag{2.30}
$$

where E_L and E_T are the longitudinal and transverse elastic moduli, respectively, G_{LT} is the shear modulus, and $v_{LT} = v_{TL}$ is the Poisson's coefficient.

In case of tubular-shaped braided specimens, three elementary tests can be per-formed. Table 2.1 gives the physical parameters related to the establishment of stress-state-applied load relationships for the three types of tests based on the strength of materials approach. Figure 2.17a presents the stress components. First, torsion test is performed along the longitudinal axis of the braided tube, to establish directly the shear modulus. Shear behavior is decoupled from other (longitudinal and transverse) elastic constants. The relationship between strain and stress is given by

$$
\varepsilon_{lt} = \frac{1}{2G_{LT}}\sigma_{LT} \tag{2.31}
$$

Second, a tensile test is performed along the axis of the tube. This test leads to a longitudinal and transverse strain field expressed by the following relationships:

$$
\varepsilon_L = \frac{1}{E_L}\sigma_L
$$

$$
\tag{2.32}
$$

$$
\varepsilon_T = \frac{v_{LT}}{E_L}\sigma_L
$$

TABLE 2.1
External Load and Geometry Data

Type of Test	Stress	Physical Parameters
Torsion	$\sigma_{LT} = \dfrac{M_t}{I_p} r$	M_t—Torque moment I_p—Moment of inertia R—Radius of the braided tube
Tensile	$\sigma_L = \dfrac{F}{S}$	F—Tensile load S—Effective cross section of the braided tube
Internal Pressure	$\sigma_L = \dfrac{Pr}{2t}$ $\sigma_T = \dfrac{Pr}{t}$	P—Internal pressure t—Thickness of the braid

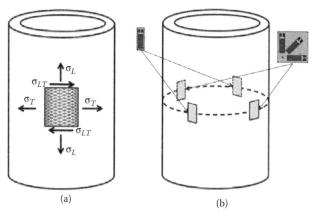

(a)   (b)

FIGURE 2.17 (a) Stress state in a braided tube subjected to tensile, internal pressure, and torsional loads. (b) Possible positioning of strain gauges and rosettes for the measurement of strain fields.

And finally, an internal pressure test is carried out on a closed tube, using water or oil under pressure, loading the specimen multiaxially, with stress fields oriented along the longitudinal and circumferential directions. In this case, the relationships between the strain- and stress-field components are written as

$$\varepsilon_l = \frac{1}{E_L}\sigma_L - \frac{v_{LT}}{E_l}\sigma_T$$

$$\varepsilon_t = \frac{1}{E_t} - \frac{v_{LT}}{E_L}\sigma_L$$

(2.33)

From these two tests, three unknown elastic constants have to be determined using the four relationships of the Equations 2.32 and 2.33. This system can lead to different results depending on the equations taken into account. To overcome this problem, it is necessary to pass through an optimization step, for example, using the least square method. The relationships between strain and stress lead to a linear system of equations written under a matrix form:

$$(\varepsilon) = [A](E)$$

$$
\begin{pmatrix} \varepsilon_L^t \\ \varepsilon_T^t \\ \varepsilon_L^p \\ \varepsilon_T^p \end{pmatrix} = \begin{bmatrix} \sigma_L^t & 0 & 0 \\ 0 & \sigma_L^t & 0 \\ \sigma_L^p & \sigma_T^p & 0 \\ 0 & \sigma_L^p & \sigma_T^p \end{bmatrix} \begin{pmatrix} \dfrac{1}{E_L} \\ \dfrac{v_{LT}}{E_L} \\ \dfrac{1}{E_T} \end{pmatrix} \qquad (2.34)
$$

where indices "t" and "p" refer to "tensile" and "pressure" tests, respectively. Generally, the vector (E) of the elastic constants is obtained by multiplying both sides of Equation 2.34 by the inverse of the matrix $[A]$:

$$[A]^{-1}(\varepsilon) = [A]^{-1}[A](E) \Rightarrow [A]^{-1}(\varepsilon) = (E) \qquad (2.35)$$

Given that the matrix $[A]$ is not a square matrix and it cannot be inversed, the resolution of the equation system can be brought to a linear approximation problem:

$$(\varepsilon) = [A](E) + (v) \qquad (2.36)$$

where (v) is the residue vector. To get the right values for the terms of the vector (E) the vector (v) has to be minimized. The use of the least square method aims at minimizing the Euclidean norm of the residue vector $\|\varepsilon - AE\|^2$. The pseudo-inverse of Moore-Pensore A^+ generally satisfies this condition, and thus the vector (E) is best estimated with

$$(E) = [A]^+(\varepsilon) \qquad (2.37)$$

The pseudo-inverse of Moore-Penrose A^+ of a matrix $[A]$ having m lines and n columns is defined as the only matrix with n lines and m columns satisfying the following conditions:

$$
\begin{aligned}
AA^+A &= A \\
A^+AA^+ &= A^+ \\
(AA^+)^T &= AA^+ \\
(A^+A)^T &= A^+A
\end{aligned}
\qquad (2.38)
$$

where T is the transpose operator. The matrix $[A]^+$ can be established using any formal mathematical software.

To measure the development of strain field in case of tubular-shaped specimens, the use of classical extensometry is adapted. Strain gauges can be bonded following the main directions. In case of tensile tests, the main deformation axis coincides with the longitudinal axis of the tube, whereas the main deformation direction can be considered at 45°. In the case of a torsional test, the strain state in the tube when subjected to an internal pressure is the superposition of longitudinal strain, similar to the tensile, and transverse circumferential strain. Therefore, the use of gauge rosettes is recommended, allowing the simultaneous measurement of different strains (Figure 2.17b). The length of the gauges has to be superior to the size of the elementary mesh, in order to avoid heterogeneity problems. It is also recommended to measure the strain field on at least two spots, sufficiently far from the clamped ends of the specimen, in order to avoid edge effects. The value of the strain given by a gauge situated in the direction α with respect to transverse axis is given by the relationship

$$\varepsilon_\alpha = \frac{\varepsilon_L + \varepsilon_T}{2} + \frac{\varepsilon_L - \varepsilon_T}{2}\cos 2\alpha + \frac{\gamma_{LT}}{2}\sin 2\alpha \qquad (2.39)$$

For an angle equal to 45°, the above relationship allows the shear strain to be estimated:

$$\varepsilon_{45} = \frac{\varepsilon_L + \varepsilon_T}{2} = \frac{\gamma_{LT}}{2} \qquad (2.40)$$

Furthermore, the measurement of the applied loads requires the use of adapted load, torque, and pressure cells.

Figure 2.18 shows an example of tubular composite specimen prepared for testing under the above-described loading conditions (Guyader, 2012). Fastening plates and devices are bonded to the tube with epoxy resin, in order to withstand and transfer the applied loads. The specimen has a length of 600 mm, an internal diameter of

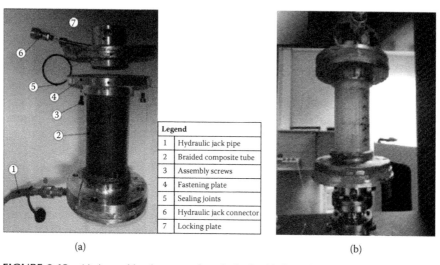

Legend	
1	Hydraulic jack pipe
2	Braided composite tube
3	Assembly screws
4	Fastening plate
5	Sealing joints
6	Hydraulic jack connector
7	Locking plate

(a)   (b)

FIGURE 2.18 (a) Assembly elements of a tubular braided specimen for testing purposes. (b) Tubular specimen installed on a universal testing machine for tensile and torsional test. (From Guyader, G., *Study of braiding conditions of high performance textile architectures: Assessment of elastic properties of 2D and 3D composite shells.* PhD thesis, University Claude Bernard Lyon 1, France, 2012. In French.)

TABLE 2.2
Mechanical Properties of the Fibers

Property	Average Value
Glass Content (%)	61.5
Tensile Modulus (GPa)	81.5
Tensile Strength (MPa)	2290
Elongation (%)	2.8

TABLE 2.3
Mechanical Properties of the Resin

Property	Average Value
Elastic Modulus (MPa)	3000
Bending Strength (MPa)	105
Tensile Strength (MPa)	80
Elongation (%)	6

100 mm, and a thickness of 10 mm. The braid reinforcement of the composite tube is obtained by a 3D braiding process, configured as a taffeta with 160 weft and 160 warp bundles. During the braiding process, horn-gears were placed on 5 levels, leading to a braid of 6 layers interlocked two by two. In the center of the horn-gear bundles were introduced to provide a longitudinal reinforcement to the braid. The tow was made of E-glass 2400 tex roving. Filaments had a diameter of 17 μm. The resin used was commercially available. Tables 2.2 and 2.3 show the properties of fiber and resin, respectively.

Specimens were produced by resin transfer molding. The braid was placed in a double-walled aluminum alloy tubular mould, and the resin was injected at a pressure of 35 bar at one end of the mold, while the other end was slightly depressurized. After setting, the composite tubes were cut to the testing dimensions. Elastic constants have been determined for several braiding angles ranging from 35° to 65°.

Considering the failure behavior of braided composite under different loading conditions, such as internal pressure or tensile load, the use of relatively large braided tubes (10–20 cm of diameter) can be unsuitable mainly for experimental reasons. In case of a tensile test, the load transfer between the fastening spares and the braid tube is made by shear at the bonded interface. To withstand the applied tensile load until failure, the area of bonded surface has to be dimensioned accordingly and can lead to excessive sized tubular specimens. When loaded under internal pressure, the circumferential stress in the braided tube is directly proportional to its radius. Therefore, to create a near-to-failure stress state in the braid, high-diameter tubes are required. Consequently, for the study of the failure behavior, tests on flat braids or coupons are recommended.

2.9 INFLUENCE OF STRUCTURE ON PROPERTIES

The influence of braid structure on the intrinsic mechanical properties has been reported in the work of Guyader (2012). Tests shown in Figure 2.18 were performed to study the influence of the braid structure on different elastic constants. First of all, the values from the measurements were directly obtained and then compared to those corrected by the least square method. Table 2.4 summarizes the obtained values and shows that it is possible to refine the rough experimental values by the least square method.

A simple way to obtain a first estimate of the elastic constants in longitudinal and transverse directions was to project the fibers volume fraction on the considered axes. Furthermore, rule of mixtures from composite mechanics was applied considering a global volume fraction of fibers of 50%, that corresponds to the ratio used during the manufacturing process of the composite tube. Table 2.5 summarizes the

TABLE 2.4
Elastic Constants of the Braided Composite as a Function of the Braiding Angle

		Braiding Angles of the Reinforcement					
		35°	40°	45°	55°	60°	65°
Longitudinal Elastic	Tensile test	19.51	20.6	19.95	18.72	18.57	21.28
Modulus E_L (GPa)	Less square method	20.28	20.97	20.28	18.64	18.65	20.81
Transverse Elastic	Internal pressure	8.76	9.9	11.54	15.25	19.08	23.41
Modulus E_T (GPa)	Less square method	7.91	8.83	10.05	13.3	16.2	20.26
Shear Modulus G_{LT} (GPa)	Torsion test	4.93	5.25	5.89	6.69	6.65	7.17
Poisson's Coefficient	Less square method	0.41	0.41	0.41	0.43	0.44	0.45

TABLE 2.5
Comparison of Experimental and Theoretical Results Obtained by the Rule of Mixtures

		35°	40°	45°	55°	60°	65°
Longitudinal Elastic	Experimental	20.2	20.97	20.3	18.64	18.65	20.81
Modulus E_L (GPa)	Law of mixtures	24.28	24.15	23.63	22.48	21.88	21.18
Transverse Elastic	Experimental	7.91	8.83	10.05	13.3	16.2	20.26
Modulus E_T (GPa)	Law of mixtures	9.97	11.2	12.25	14.18	15.05	15.75

theoretical values, which overestimate the experimental data, but confirms the same tendency of the influence of braiding angle on the elastic constants.

Figures 2.19 and 2.20 show the change of elastic constants as a function of the braiding angle. The longitudinal elastic modulus E_{LT} is slightly influenced by the variation of the braiding angle. This is explained by the braiding configuration, in which a large number of bundles are used to reinforce longitudinally the composite (320 longitudinal bundles, 160 weft and 160 warp off-axis bundles). In the case of braids where bundles are oriented only off-axis from the loading direction, the elastic modulus is more sensitive to the braid angle and to the crimp of the bundles. In this case, it is possible that the tensile and compressive properties of the braided composite are lower than those of 2D laminates (Mouritz et al., 1999).

Transverse and shear modulus are more sensitive to the variation of the braid angle; a linear estimate of these constants as a function of the braiding angle fits with sufficient accuracy to the experimental values.

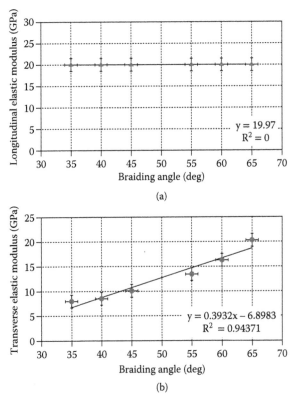

FIGURE 2.19 Influence of braiding angle variation on the stiffness of braided composite tubes. (a) Effect on the longitudinal elastic modulus. (b) Effect on the transverse elastic modulus. (From Guyader, G., *Study of braiding conditions of high performance textile architectures: Assessment of elastic properties of 2D and 3D composite shells.* PhD thesis, University Claude Bernard Lyon 1, France, 2012. In French.)

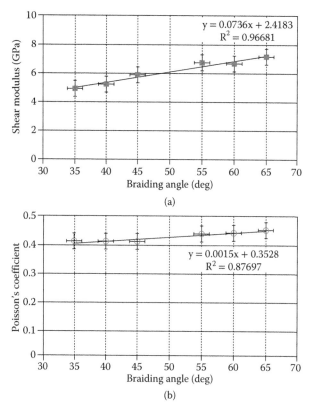

FIGURE 2.20 Influence of braiding angle variation on the mechanical properties of braided composite tubes: (a) Effect on shear modulus. (b) Effect on the Poisson's ratio. (From Guyader, G., *Study of braiding conditions of high performance textile architectures: Assessment of elastic properties of 2D and 3D composite shells*. PhD thesis, University Claude Bernard Lyon 1, France, 2012. In French.)

Another aspect of the influence of structure on properties can be related to the surface texture, when the braid is used as an internal reinforcement as reinforcement bar or equivalent. In the works of Cunha et al. (2014), the use of braided rods with braided yarns in the outer layer of the braid structure and the adjustment of the process parameters of the braiding operation lead to a tailored surface roughness of the rod. Consequently, the interface behavior between the rod and the matrix (cimentitious) is highly improved, leading to a very effective use of the reinforcement, utilizing fully the tensile strength of the braided rods.

2.10 CONCLUSIONS

This chapter presented the main parameters of the production of braids used in composite industry. There is a strong interaction between the process parameters and the properties of the final product, braid preform, and the composite. Consequently, the modeling of the parameters of the manufacturing process of braids is necessary to

achieve adapted production rates in correlation with industrial production requirements. The theoretical base developed in this chapter, corresponding to transient or
steady-state production phases, has to be implemented in computer-assisted production software to allow production control. The control of the properties of the preform
or the final composite is necessary and can be made using nondestructive investiga-
tions. Mechanical properties can be evaluated using classical extensometry or image
correlation. Nevertheless, the data obtained by these techniques are gathered observ-
ing the external/internal surface of the braid, and therefore, in case of 3D interlock
composites, where several layers are interlaced, the behavior at the interface or in the
core of the braid is not analyzed with sufficient accuracy. Consequently, it is necessary to develop new experimental analysis techniques or combined experimental-
numerical methods better adapted to the analysis of the core behavior.

REFERENCES

Brown, R. and Crow Jr., E. (1992). Automatic through-the-thickness braiding. In *Proceedings of the 37th International SAMPE Symposium*, 832–842 (1992).

Calme, O., Bigaud, D., Jones, S., and Hamelin, P. (2006). Analytical evaluation of stress state in braided orthotropic composite cylinders under lateral compression. *Composites Science and Technology*, 66(15):3040–3052.

Chiu, C. (1999). Crush-failure modes of 2D triaxially braided hybrid composite tubes. *Composites Science and Technology*, 59(11):1713–1723.

Cox, B., Dadkhah, M., Morris, W., and Flintoff, J. (1994). Failure mechanisms of 3D woven composites in tension, compression, and bending. *Acta Metallurgica et Materialia*, 42(12):3967–3984.

Cunha, F., Rana, S., Fangueiro, R., and Vasconcelos, G. (2014). Excellent bonding behaviour of novel surface-tailored fibre composite rods with cementitious matrix. *Bulletin of Materials Science*, 37(5):1013–1016.

de Oliveira Simões, J. and Marques, A. (2001). Determination of stiffness properties of braided composites for the design of a hip prosthesis. *Composites Part A: Applied Science and Manufacturing*, 32(5):655–662.

Du, G. and Popper, P. (1994). Analysis of a circular braiding process for complex shapes. *Journal of the Textile Institute*, 85(3):316–325

Falzon, P. J. and Herszberg, I. (1998). Mechanical performance of 2-D braided carbon/epoxy composites. *Composites Science and Technology*, 58(2):253–265.

Guyader, G. (2012). *Study of braiding conditions of high performance textile architectures: Assessment of elastic properties of 2D and 3D composite shells*. PhD thesis, Universit Claude Bernard Lyon 1. In French.

Guyader, G., Gabor, A., and Hamelin, P. (2013). Analysis of 2D and 3D circular braiding processes: Modeling the interaction between the process parameters and the pre-form architecture. *Mechanism and Machine Theory*, 69:90–104.

Hamada, H., Nakatani, T., Nakai, A., and Kameo, K. (1999). The crushing performance of a braided I-beam. *Composites Science and Technology*, 59(12):1881–1890.

Kessels, J. and Akkerman, R. (2002). Prediction of the yarn trajectories on complex braided preforms. *Composites Part A: Applied Science and Manufacturing*, 33(8):1073–1081.

Ko, F. (1987). Braiding. In *Engineering Materials Handbook*. ASM International, Netherlands.

Kohlman, L. W., Bail, J. L., Roberts, G. D., Salem, J. A., Martin, R. E., and Binienda, W. K. (2012). A notched coupon approach for tensile testing of braided composites. *Composites Part A: Applied Science and Manufacturing*, 43(10):1680–1688.

Lee, S., editor (1993). *Handbook of Composite Reinforcement*. VCH Publishers, Germany.

Leung, C. K., Melenka, G. W., Nobes, D. S., and Carey, J. P. (2013). The effect on elastic modulus of rigid-matrix tubular composite braid radius and braid angle change under tensile loading. *Composite Structures*, 100:135–143.

Michaeli, W., Rosenbaum, U., and Jehrke, M. (1990). Processing strategy for braiding of complex-shaped parts based on a mathematical process description. *Composites Manufacturing*, 1:243–249.

Miravete, A., editor (1999). *3-D Textiel Reinforcements in Composite Materials*. Woodhead Publishing, UK.

Mouritz, A., Bannister, M., Falzon, P., and Leong, K. (1999). Review of applications for advanced three-dimensional fibre textile composites. *Composites Part A: Applied Science and Manufacturing*, 30(12):1445–1461.

Pickett, A. and Fouinneteau, M. (2006). Material characterisation and calibration of a meso-mechanical damage model for braid reinforced composites. *Composites Part A: Applied Science and Manufacturing*, 37(2):368–377.

Potluri, P., Manan, A., Francke, M., and Day, R. (2006). Flexural and torsional behaviour of biaxial and triaxial braided composite structures. *Composite Structures*, 75(1-4):377–386.

Priem, C., Othman, R., Rozycki, P., and Guillon, D. (2014). Experimental investigation of the crash energy absorption of 2.5D-braided thermoplastic composite tubes. *Composite Structures*, 116:814–826.

Stover, E., Marck, W., Marfowitz, I., and Mueller, W. (1971). Preparation of an omniweave-reinforced carbon/carbon cylinder as a candidate for evaluation in the advanced heat shield screening program. Technical Report, AFML-TR-70-283.

Subhash, G., Sulibhavi, S., and Zikry, M. A. (2001). Influence of strain-rate on the uniaxial compressive behavior of 2-D braided textile composites. *Composites Part A: Applied Science and Manufacturing*, 32(11):1583–1591.

Sutcliffe, M., Monroy Aceves, C., Stronge, W., Choudhry, R., and Scott, A. (2012). Moderate speed impact damage to 2D-braided glasscarbon composites. *Composite Structures*, 94(5):1781–1792.

Tada, M., Uozumi, T., Nakai, A., and Hamada, H. (2001). Structure and machine braiding procedure of coupled square braids with various cross sections. *Composites Part A: Applied Science and Manufacturing*, 32(10):1485–1489.

Zhang, Q., Beale, D., and Broughton, R. (1999a). Analysis of circular braiding process, part 1: Theoretical investigation of kinematics of the circular braiding process. *Journal of Manufacturing Science and Engineering*, 121(3):345–350.

Zhang, Q., Beale, D., and Broughton, R. (1999b). Analysis of circular braiding process. part 2: Mechanics analysis of the circular braiding process. *Journal of Manufacturing Science and Engineering*, 121(3):351–357.

Zsigmond, B. (2005). *Modeling of Braided Fiber Reinforced Composites Crosslinked by Electron Beam*. PhD thesis, Technical University Budapest, Hungary.

3 Analysis of Braided Structures and Properties

Bohong Gu

CONTENTS

ABSTRACT

This chapter reports the structure characterizations and mechanical behaviors of two-dimensional (2D) and three-dimensional (3D) braided preforms and composite materials. The microstructure modeling strategies from both braided preforms and unit cells are presented. The uniaxial tensile/compressive stiffness and strength of braided preforms and composites are derived from the microstructure models and unit cell model. Especially, the dynamic mechanical behaviors, that is, the mechanical behaviors under impact loadings and cyclic loadings are reported in this chapter. Some design examples are given for engineering applications.

3.1   MODELING OF BRAIDED STRUCTURES

3.1.1   TWO-DIMENSIONAL BRAIDS

Braided preforms and composite materials have been widely applied to engineering structures, especially complex shape structures. The advantages of braided structures are the higher mechanical behaviors along thickness directions than 3D woven or knitted composite materials. Furthermore, braided composite materials have a higher degree of isotropic features than woven or knitted composite materials. Usually, there are two types of braided preforms, that is, 2D and 3D braided structures [1]. Figure 3.1 shows 2D braided structures. Figure 3.1a shows the braid

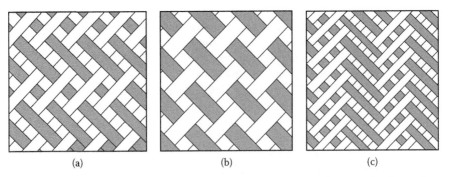

(a)   (b)   (c)

FIGURE 3.1   Geometric configuration of (a) regular or plain braid weave, (b) diamond or basket braid weave, and (c) triaxial or Hercules braid weave.

with a 2/2 repeat of the intersecting pattern, and it is known as a regular, plain, or standard braid. Figure 3.1b gives a diamond or basket braid, which is characterized by a 1/1 repeating pattern. Figure 3.1c is a regular braid with longitudinal in-laids, and it is also called a triaxial braid because of the three directional fiber orientations. To improve the axial tensile properties, a set of axially oriented yarns can be incorporated into the braided structure. These yarns are supplied from a set of stationary packages and not truly interwoven with the bias yarns, which are themselves mutually interwoven, as shown in Figure 3.2.

There are certain significant similarities and differences between woven and braided fabrics. Both fabrics utilize two sets of yarns; these are the warp and weft yarns in weaving and the yarns moving in clockwise and counterclockwise directions in circular braiding. The angle of interlacing between the two sets of yarns is 90° in orthogonal woven fabrics and less than 90° in braided fabrics [2]. Usually, the 2D braided performs or composite materials are produced from a two-step braiding technique. The structure of a two-step braid involves a large number of axial yarns, which can be arranged in essentially any shape, including I beams, box beams, circular tubes, and so on. The braiders (braiding yarns) are arranged around the perimeter of the axial array as shown in Figure 3.3. In the braiding process, the braiders move through the axial array in two sequential steps. In the first step, the braiders all move in one diagonal line but in alternating directions. In the second step, they move along the other diagonal line.

The surface of the two-step braids is shown in Figure 3.4 [1]. The pattern is characterized by the angular orientation of braider yarns, θ_b, and the pitch length, h. Compared with other 3D braiding processes, two-step braiding has several distinct advantages. A relatively simple sequence of braider motions can form a wide range of shapes. During each step of the process, all the braiders are simultaneously outside the axial array, and thus it is possible to add various inserts to the structure or even rearrange the axial array geometry to change the preform cross section, such as I beams, box beams, circular tubes, and so on. Furthermore, a high level of fiber packing (up to 78%) can be achieved in composites reinforced with two-step braided preforms [3].

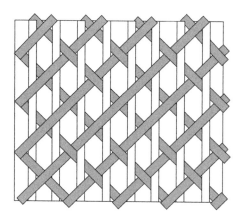

FIGURE 3.2 Braided structure with axial in-laid yarn.

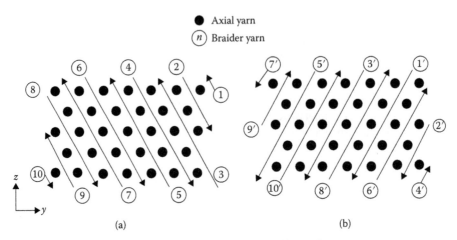

FIGURE 3.3   Two-step braiding process: (a) step 1 and (b) step 2.

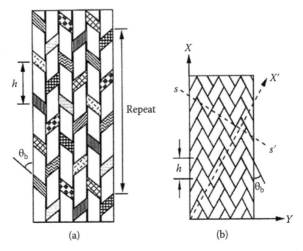

FIGURE 3.4   Schematic surfaces of (a) a two-step braid and (b) a four-step braid.

The geometric model for two-step braided composites is established based on the knowledge of the cross sections of axial yarns at the center, side, and corner positions, after matrix impregnation (Figure 3.5). The cross section of braiding yarns is assumed to be rectangular. Referring to Figure 3.5, the yarn dimensions can be expressed in terms of yarn properties and processing parameters [3].

$$S_a = \left[\frac{\lambda_a}{\gamma_a v_a \sin(2\theta)} \right]^{1/2} \tag{3.1}$$

$$S_n = \frac{0.5\lambda_a}{2S_a \gamma_a v_a \cos\theta} \tag{3.2}$$

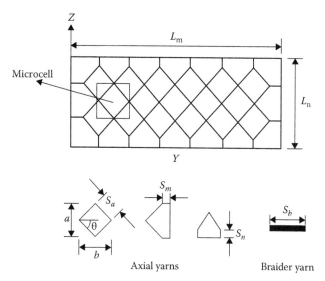

Axial yarns   Braider yarn

FIGURE 3.5   Cross sections of a two-step braided composite and the constituent yarns.

$$S_m = \frac{0.5\lambda_a}{2S_a\gamma_a v_a \sin\theta} \tag{3.3}$$

$$S_b = \left(\frac{\lambda_b}{\gamma_b v_b f_b}\right)^{1/2} \tag{3.4}$$

where n and m are the yarn row number and column number respectively; λ_a is the linear density of center axial yarns (kg/m); λ_b is the linear density of braider yarns (kg/m); γ_a and γ_b are fiber densities of axial and braider yarns, respectively (kg/m³); v_a and v_b are fiber packing fractions of axial and braider yarns, respectively; and f_b is the aspect ratio of braider yarns. Referring to Figure 3.5, the aspect ratio of axial yarns, f_a, is defined as

$$f_a = \frac{a}{b} = \tan\theta \tag{3.5}$$

Based on the relationships of Equations 3.1 through 3.5, a unit cell of the braided composite can be defined. Figure 3.6 shows the proposed unit cell geometry from the work of Du, Popper, and Chou [3]. The length of the unit cell is the length of the braid formed in one machine step, which is one half of the fabric pitch length, h. It should be noted that in the definition of a unit cell the number of columns of the braided composite in Figure 3.5 has been assumed to be very large so that the geometric edge effect can be ignored.

A key processing parameter of a unit cell is the braiding yarn orientation angle, α, which is the angle between the braider and axial yarns. The inclined braider yarn in

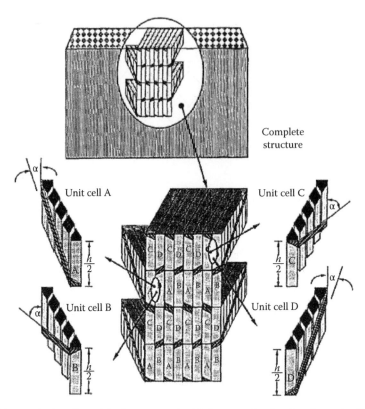

FIGURE 3.6 Unit cell model of a two-step braided composite showing four subcells.

unit cell A of Figure 3.6, for instance, is assumed to lie along the diagonal direction of the parallelepiped. The x direction coincides with the braiding axis, which forms the angle α with the braider yarn. From Figure 3.7, the braider yarn orientation is given by

$$\alpha = \tan^{-1}\left(\frac{2L_{\mathrm{p}}}{h}\right) \qquad (3.6)$$

where L_{p} is the projected length of the yarn in Figure 3.7 into the X–Z plane. From Equations 3.1 through 3.4,

$$L_{\mathrm{p}} = (n-1)\left[S_{\mathrm{a}} + \frac{S_{\mathrm{b}}f_{\mathrm{b}}}{\sin(2\theta)}\right] + 2S_n \qquad (3.7)$$

where n is the number of layers in the cross section of the preform. On the basis of the unit cell model, the volume fractions of axial yarns, braider yarns, and interyarn voids can be analyzed. The study by Du, Chou, and Popper [3] shows good agreements in cross-sectional dimension, fiber volume fraction, and braider yarn orientation angle between theoretical predictions and experimental measurements for two-step braid composites of seven columns by five layers.

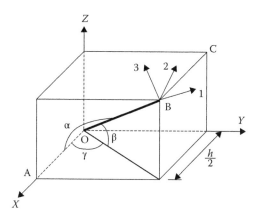

FIGURE 3.7 The braider yarn in the unit cell A of Figure 3.6.

3.1.2 THREE-DIMENSIONAL BRAIDS

Three-dimensional braided preforms and composite materials are usually obtained from Cartesian braided techniques, in which four-step rectangular braiding is the most common method.

Figure 3.8 shows the process of braiding a textile preform with a rectangular cross section. The braiding yarns arrange into a rectangle with rows and columns. This part of the yarn is called the main participant yarns, denoted as $m \times n$. It is noted that m means the row and n means the column. At the outside of the main yarns, the braiding yarns are arranged at intervals, which means the braiding yarns are out of the rectangular section. This part of the braiding yarn is called the side yarns. There is no special rule for the arrangement of side yarns. However, it is required that the number of yarns in every row (or column) made of main yarns and side yarns should be equal to the other row (or column). During the braiding process, due to the moving of the yarns, at a certain occasion, every braiding yarn could be the main yarn; and at the other certain occasions, the braiding yarn could become the side yarn. During the movement of the yarn-carrier device, when the yarn moves by the horizontal direction the yarn moves only by one yarn position. When the yarn moves in the longitudinal direction, the yarn also moves by only one yarn position. Therefore, this cell is called the 1×1 structure, and this is the simplest structure, which has been used widely. Furthermore, there are structures such as 1×2, 1×3, 2×3, and so on. The first number means the number of yarn positions by which the yarn moves in the first and third step. The second number means the number of yarn positions by which the yarn moves in the second and fourth step.

For a rectangular preform consisting of n rows and m columns of braiding yarns, the total number of yarns with a 1×1 rectangular pattern is

$$N = m \times n + m + n \tag{3.8}$$

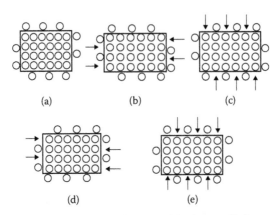

FIGURE 3.8 Four-step rectangular braiding: (a) original state, (b) first step, (c) second step, (d) third step, and (e) fourth step.

The total N braiding yarns can be divided into G groups according to the sequence of braiding. Within a group, the motion of individual yarns follows the same path. The number of yarn groups in an $m \times n$ rectangular braided preform is given by

$$G = m \times n/\lambda_{mn} \qquad (3.9)$$

where λ_{mn} is the least common multiple of m and n.

Figure 3.9 shows the surface of four-step braided preforms, in which h is the preform pitch length that is formed in one machine cycle and α is the surface angle, which is directly measurable at the surface. These two parameters are essentially important in characterizing the structure of the preforms.

Assume a circular yarn cross section. An idealized model to represent the inside structure of a 1×1 four-step rectangular preform is shown in Figure 3.9. It shows the cross section cut longitudinally at 45° angle with the preform surface. Because of the symmetrical nature of the structure, the cross sections $ABCD$ and $CDEF$ are the same. It indicates that the yarns inside the structure are inclined in four directions. Two of these directions are parallel to the X–Z plane, and the other two are parallel to the Y–Z plane. This figure also shows that yarns inclined in the same direction form layers in the structure that are alternately packed with layers formed by yarns inclined in the conjugate direction in a parallel plane. The orientation angles of the yarns inclined in all directions are the same and denoted as γ. The following equation is deduced:

$$h \tan \gamma = 4d \qquad (3.10)$$

Under jamming condition, the pitch reaches its minimum value:

$$h_{\min} = \frac{2d}{\tan \gamma} \sqrt{2 + \tan^2 \gamma} \qquad (3.11)$$

From Equations 3.10 and 3.11, we can find that the minimum pitch length $h_{\min} = 2.8d$ and the maximum orientation angle is $\gamma_{\max} = 55°$.

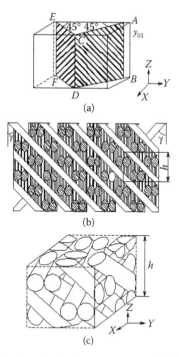

FIGURE 3.9 Geometrical model for the four-step rectangular braided preform: (a) cut longitudinally at 45° angle, (b) cross section, and (c) unit cell structure.

Because the surface yarns move a shorter distance in the transverse plane and the same distance in the axial direction as the inner yarns in one machine cycle, their orientation angle is smaller than that of the inner yarns. The surface yarn orientation angle β and its relation with γ and α can be best represented as shown in Figure 3.10 by assuming that a surface yarn is composed of two straight segments. The surface angle α is actually the projection of β on the preform surface.

Based on the geometric relations between inner yarn orientation angles, the surface yarn orientation angle and surface angle for a rectangular preform can be written as follows:

$$\tan \alpha = \tan \gamma / 2\sqrt{2} = \tan \beta / \sqrt{2} \qquad (3.12)$$

Periodically, every braiding yarn appears in the inner and surface parts of the preform. The cross section of the preform is composed of yarns from inner and surface (including corner) parts with different orientation angles. For an $m \times n$ rectangular braided preform, the number of inner yarns is $m \times n - m - n$ and the number of surface yarns is $2(m + n)$. If an analysis is based merely on the inner structure and neglects the surface effects, then only a higher percentage of the inner yarn number with respect to the surface yarn number will give accurate results.

The contour size of the preform can be defined by the width of the sides, which can be obtained according to its inside structure. From Figure 3.9, it is shown that

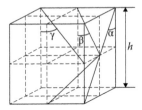

FIGURE 3.10 Relation between the angles of a rectangular preform.

the yarns oriented in the same direction form layers inclined to the preform sides at 45° angles. The neighboring layers are in contact with each other to meet the constraint requirement. Then, the width of a rectangular preform having k yarns is as follows:

$$W_k = d \times \left(\sqrt{2}k + 1\right) \tag{3.13}$$

From Figure 3.9, the unit volume of the rectangular preform, U, and the total volume of the yarns in the unit, Y, can be derived as follows:

$$U = h^3 \tan^2 \gamma$$

$$Y = \frac{8\pi(d/2)^2 h}{\cos^2 \gamma}$$

The yarn volume fraction of the four-step braided preform is defined by the total yarn volume divided by the volume of the structure containing these yarns. The fiber volume fraction, V_f, can then be obtained by multiplying the yarn volume fraction by the fiber packing fraction, κ. Thus,

$$V_f = \frac{\pi\kappa}{8h}\sqrt{h^2 + 16d^2} \tag{3.14}$$

The fiber volume fraction in Equation 3.14 is derived on the basis of a unit taken from the inner part of the structure. When the preform is composed of a small number of yarns, surface effect should be considered. An overall fiber volume fraction based on the whole preform structure is derived by taking a length of the preform that includes both inner and surface yarns.

$$V_f = \frac{\pi\kappa}{h\left(6.828 + 1.172c_i\right)}$$
$$\left[c_i\sqrt{h^2 + 16d^2} + \left(1 - c_i\right)\sqrt{h^2 + 4d^2}\right] \tag{3.15}$$

where c_i is the number fraction of the inner yarns and κ is the fiber packing fraction. It can be seen that when $c_i = 1$, Equation 3.15 is identical to Equation 3.14.

When the braided structure reaches its jamming condition, the fiber volume fraction attains its maximum value. By Equation 3.14 and assuming $\kappa = 1$, the maximum fiber volume faction is 0.685 when $\gamma = 55°$.

A refined geometrical model of the braided preforms can further divide the cross section of a rectangular braided preform into three regions, that is, inner, surface, and corner regions [4]. For an $m \times n$ braided preform, the volumes of each region in terms of the whole volume of the preform are given in Table 3.1. Detailed model development and derivation of the corresponding equations can be found in the literature [5,6].

The fiber volume fraction in the corresponding region can be calculated as follows:

$$V_{if} = \frac{\pi\kappa\sqrt{3}}{8}$$

$$V_{sf} = \frac{\pi\kappa\sqrt{3}}{24}\left(1 + \frac{3\cos\gamma}{\cos\theta}\right) \qquad (3.16)$$

$$V_{cf} = \frac{\pi\kappa3\sqrt{3}\cos\gamma}{17\cos\beta}$$

where V_{if}, V_{sf}, and V_{cf} are the fiber volume fractions in the inner, surface, and corner regions, respectively. The angles θ, β, and γ are the orientation angles of yarn in the surface, corner, and inner regions, respectively.

TABLE 3.1

Volume Proportion of Each Region to Whole Structure

	m,n are Even	m or n is Odd
Inner region	$V_i = \dfrac{2(m-1)(n-1)+2}{2mn+m+n}$	$V_i = \dfrac{2(m-1)(n-1)}{2mn+m+n}$
Surface region	$V_s = \dfrac{3(m+n-4)}{2mn+m+m}$	$V_s = \dfrac{3(m+n-2)}{2mn+m+m}$
Corner region	$V_c = \dfrac{8}{2mn+m+m}$	$V_c = \dfrac{4}{2mn+m+m}$

Source: Chen et al., *Journal of China Textile University (Eng. Ed.)*, 14(3), 8–13, 1997.

The total fiber volume fraction is the summation of the fractions of the three regions:

$$V_f = V_i V_{if} + V_s V_{sf} + V_c V_{cf} \tag{3.17}$$

3.2 MODELING OF MECHANICAL PROPERTIES OF THREE-DIMENSIONAL BRAIDED PREFORMS

In this section, the tensile curve of a 3D braided preform within the whole tensile strain is predicted on the basis of mathematical description of microstructure and energy conservation law. From the tensile curve, the strength of the 3D braided preform could also be obtained. A comparison of the theoretical and experimental tensile curves of the 3D braided preform indicates that the method in this chapter could predict the uniaxial tensile curve correctly in a simple way.

3.2.1 MATHEMATICAL DESCRIPTION OF MICROSTRUCTURES OF THREE-DIMENSIONAL BRAIDED PREFORMS

To get spatial configurations of braiding yarns in a braided preform, trace yarn was used in manufacturing a four-step 3D braided preform. In the experiment, the braiding yarns were plied cotton yarn and the trace yarn was a colored plied cotton yarn. In each yarn group, one trace yarn was used in braiding. According to the simulation results of Wang and Sun [7], the spatial configuration of all yarns in one yarn group is the same and the only difference is the initial position in the axial direction. When the spatial configuration of traced yarn is obtained, the configuration of other yarns in the same yarn group can also be obtained. To get the spatial configuration of trace yarn, it is necessary to locate the coordinates of center points in a series of cross sections of the trace yarn. If the interval between two neighboring points is small enough, the line passing these points can converge to the real spatial configuration of trace yarn in braided preform. Because the braiding yarns appear straight inside the braided preform, bending and then change to other directions occur only at the surface. We could cut the braided preform at the transition point of trace yarn on braided preform surface to obtain cross section of the braided preform. Then, the straight line between the two transition points could represent the trace yarn configuration in the interior of braided preform. In a machine cycle (i.e., all yarn carriers returning to their original positions), the spatial configuration of trace yarn could be expressed by linking a series of straight lines.

Assume that the coordinates of center points of trace yarn in two neighboring cross sections are (x_i, y_i, z_i) and $(x_{i+1}, y_{i+1}, z_{i+1})$, respectively; then, the equation of straight line between the two points (i.e., the equation of the ith section of trace yarn) [8] is as follows:

$$\begin{cases} x_i = M_{i1}(t - b_i) + M_{i2} \\ y_i = N_{i1}(t - b_i) + N_{i2} \qquad b_i \le t \le b_{i+1} \\ z_i = t \end{cases} \tag{3.18}$$

where

$$M_{i1} = \frac{x_{i+1} - x_i}{z_{i+1} - z_i} \qquad M_{i2} = x_i$$

$$N_{i1} = \frac{y_{i+1} - y_i}{z_{i+1} - z_i} \qquad N_{i2} = y_i \qquad (3.19)$$

$$b_i = z_i$$

Because the other yarns in one yarn group have identical spatial configurations at a fixed space coordinate, the only difference among them is the initial position along the longitudinal direction of braided preform. Then, these yarns could be expressed with equations of the same form and coefficient as Equations 3.18 and 3.19. This phenomenon is like that of periodic functions of different phases.

We can assume the following:

1. There are M_r yarns in one yarn group.
2. Yarns are numbered using j ($j = 0, 1, \ldots, M_r - 1$).
3. The 0th yarn is datum yarn.

Then, the initial position of jth yarn relative to 0th yarn is as follows:

$$s_j = j \cdot h / M_r \qquad (3.20)$$

The ith segment in jth yarn in one yarn group could be expressed as follows:

$$\begin{cases} x_{ji} = M_{i1}(t + s_j - b_i) + M_{i2} \\ y_{ji} = N_{i1}(t + s_j - b_i) + N_{i2} \quad b_i \le t \le b_{i+1} \\ z_{ji} = t \end{cases} \qquad (3.21)$$

where the parameters in Equation 3.21 are the same as those in Equation 3.17. This is the equation to describe the spatial configuration of every yarn in one yarn group.

3.2.2 RELATIONSHIPS BETWEEN UNIAXIAL TENSILE STRAIN OF BRAIDED PREFORM AND BRAIDING YARNS

Figure 3.11 shows the straight line of braiding yarn segment between two continuous cross sections, where θ is the angle between $o–z$ axis and straight line of braiding yarn section. Figure 3.12 shows the xoz plane of Figure 3.11. In Figures 3.11 and 3.12, oz is the axis direction of braided preform.

3.2.2.1 Small Strain Theory

For small strains, changes in the geometry are such that second-order effects can be neglected, and there is no lateral contraction of the specimen under uniaxial tensile stress. Consider the fiber segment op in Figure 3.12; the upper end of the yarn p

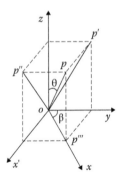

FIGURE 3.11 Braiding yarn line between two continuous cross sections in a perform.

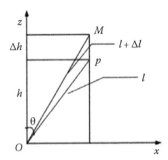

FIGURE 3.12 The *xoz* plane of Figure 3.11.

moves vertically to M when the elongation of braided preform is Δh. If we assume that the lengths of braided preform and yarn segment between the two continuous cross sections are h and l, respectively, before tension, then the tensile strain of braided preform is $\varepsilon_b = \Delta h/h$ and that of braiding yarn is $\varepsilon_y = \Delta l/l$.

The relationship between braided preform strain and braiding yarn strain is as follows:

$$\varepsilon_y^2 + 2\varepsilon_y - \cos^2 \theta \left(\varepsilon_b^2 + 2\varepsilon_b \right) = 0$$
$$\Rightarrow \varepsilon_y = \sqrt{1 + \left(\varepsilon_b^2 + 2\varepsilon_b \right)\cos^2 \theta} - 1 \tag{3.22}$$

3.2.2.2 Large Strain Theory (General Theory)

For large strains, the change of angle θ and the lateral contraction should be considered, leading to the theory of Poisson's ratio, ν, which is defined as the ratio of lateral strain to strain in the direction of applied stress. Under uniaxial tension, the position of braiding yarn, *op*, changes to *oN* in Figure 3.13.

In Figure 3.13, $\Delta r = r - r'$, and the Poisson's ratio is

$$\nu = \frac{\Delta r}{r} / \left(\Delta h/h \right) \quad \Rightarrow \quad \Delta r = \nu \cdot \Delta h \cdot \tan \theta \tag{3.23}$$

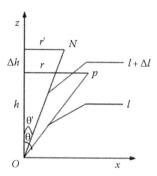

FIGURE 3.13 Deformation of braiding yarn in preform when uniaxial large strain is applied.

The length of braiding yarn before tension is $l = h/\cos\theta$.
After tension,

$$l + \Delta l = \left[(h + \Delta h)^2 + (h \cdot \tan\theta - v \cdot \Delta h \cdot \tan\theta)^2 \right]^{1/2} \tag{3.24}$$

Yarn strain is

$$\varepsilon_y = \frac{\Delta l}{l} = \frac{\left[(h + \Delta h)^2 + (h \cdot \tan\theta - v \cdot \Delta h \cdot \tan\theta)^2 \right]^{1/2}}{h/\cos\theta} - 1 \tag{3.25}$$

The relationship between braided preform strain and braiding yarn strain is as follows:

$$\varepsilon_y = \left[(1 + \varepsilon_b)^2 \cdot \cos^2\theta + \sin^2\theta (1 - v\theta_b)^2 \right]^{1/2} - 1 \tag{3.26}$$

3.2.2.3 Direction Cosine of Braiding Yarn Segment

The direction cosine of the ith segment of braiding yarn before tension could be determined by Equation 3.18 as follows:

$$\cos\theta = \frac{z'(t)}{\sqrt{x'^2(t) + y'^2(t) + z'^2(t)}} = \frac{1}{\sqrt{M_{i1}^2 + N_{i1}^2 + 1}} \tag{3.27}$$

The angle θ in Figures 3.12 and 3.13 will vary with the uniaxial elongation of braided preform. In the case of large strains, the angle θ becomes θ' in Figure 3.13. Then, the direction cosine is expressed as follows:

$$\begin{cases} \cos\theta' = (h + \Delta h) \Big/ \sqrt{(h + \Delta h)^2 + r'^2} \\ r/h = \cos\theta \end{cases} \tag{3.28}$$

The direction cosine of the braiding yarn segment after tension is

$$\cos\theta' = \cos\theta \Big/ \sqrt{\cos^2\theta + \sin^2\theta(1-\nu\varepsilon_b)^2 \Big/ (1+\varepsilon_b)^2} \qquad (3.29)$$

3.2.2.4 Transforming Braiding Yarn Strain to Stress

At low strains, this transformation may be accomplished by assuming a Hookean relation between braiding yarn strain and stress. Thus,

$$\sigma_y = E_y\varepsilon_y \qquad (3.30)$$

For large strains, it is necessary to take into account deviation from Hooke's law. The stress σ_y may either be read directly from the stress–strain curve or be taken from some other approximation forms of the fitted equation. In general, it can be expressed as follows:

$$\sigma_y = F(\varepsilon_y) \qquad (3.31)$$

3.2.3 RELATIONSHIPS BETWEEN STRESS OF BRAIDING YARN AND BRAIDED PREFORM

When the external uniaxial tensile force p is applied to braided preform, the uniaxial deformation is Δh, as illustrated in Figure 3.13, and the load of the jth braiding yarn is F_j, the elongation of which is ds_j.

Because the work of external uniaxial force applied to braided preform is equal to the sum of works of every braiding yarn caused by F_j,

$$p \cdot dh = \sum_{j=1}^{N} F_j \cdot ds_j \Rightarrow p = \sum_{j=1}^{N} F_j \cdot ds_j \Big/ dh \qquad (3.32)$$

Assume that the lengths of braided preform and braiding yarn before tension are h and l, respectively, and the strains of braided preform and braiding yarn are ε_b and ε_y, respectively. Then, Equation 3.32 can be transformed as follows:

$$p = \sum_{j=1}^{N} F_j \cdot \frac{l}{h} \cdot \frac{d\varepsilon_y}{d\varepsilon_b} \qquad (3.33)$$

From Equation 3.33, the stress–strain of braided preform could be obtained when stress–strain of braiding yarn, relationship between strain of braiding yarn and braided preform under uniaxial tension, and direction angle of each yarn segment are all known. Then, the whole uniaxial tensile curve within failure strain can be predicted. The flowchart for these discussions is shown in Figure 3.14.

The maximum strain failure criterion is used in the calculation. When the strain of one braiding yarn reaches maximum failure strain, this braiding yarn can no longer bear the external load. Only when all braiding yarns are broken, the braided preform can be regarded to fail completely.

3.2.4 Examples for Experimental Verifications

3.2.4.1 Microstructure of Four-Step Three-Dimensional Braided Preform

A cotton plied yarn (1057 tex) was used in the manufacture of braided preform by the four-step 1 × 1 method. Two kinds of braided preforms, namely, 6 × 4 array and 6 × 6 array of rectangular cross section, were manufactured (hereinafter referred to as 6 × 4 and 6 × 6 preforms, respectively). According to the theory of Li and El Shiekh [9], the numbers of yarn groups are two for 6 × 4 preform and six for 6 × 6 preform, respectively. In each yarn group, a colored yarn acted as the trace yarn. The sizes of cross section of 6 × 4 preform and 6 × 6 preform are 13.4 × 9.6 mm and 13.4 × 13.4 mm, respectively. The lengths of 6 × 4 preform and 6 × 6 preform in a machine cycle (i.e., complete cycle of movement of yarn carriers) are 75.6 mm and 32.8 mm, respectively. For the trace yarn of 6 × 4 preform in one group, there are 11 transition points at braided preform surface in a machine cycle, and the spatial configuration of this trace yarn could be described with 10 equations of straight segment line in 3D space. For the trace yarn of 6 × 6 preform, the number of transition points is five and four equations are enough for describing the spatial configuration of trace yarn. The cross section of braided preform was magnified and photographed with Quester 3D video microscopy. From the coordinates of a series of center points in the cross section of trace yarn, the spatial configuration could be described as discussed in Sections 3.2.4.1.1 and 3.2.4.1.2.

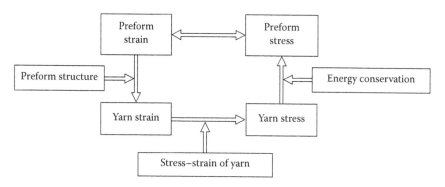

FIGURE 3.14 Flowchart of prediction of uniaxial tensile curve of braided preform.

3.2.4.1.1 Preform 6 × 4

For the 6 × 4 preform, the yarn group is two and there are 17 yarns in each group. The spatial configuration could be described with Equation 3.18. The coefficients of Equation 3.18 of the 0th yarn of the first yarn group are as follows:

$$
M = \begin{bmatrix}
-0.174 & 2.29 \\
0.940 & 0.94 \\
0.655 & 8.14 \\
-0.531 & 13.03 \\
-0.882 & 10.08 \\
-0.137 & 1.98 \\
0.860 & 0.92 \\
0.739 & 7.56 \\
-0.502 & 13.03 \\
-0.846 & 10.11
\end{bmatrix}
\qquad
N = \begin{bmatrix}
-0.490 & 7.83 \\
-0.577 & 6.96 \\
0.704 & 0.38 \\
0.631 & 5.63 \\
0.607 & 9.14 \\
0.514 & 0.90 \\
0.760 & 3.05 \\
-0.716 & 8.92 \\
-0.546 & 3.62 \\
0.780 & 0.44
\end{bmatrix}
$$

$$
b = \begin{pmatrix} 0 & 7.76 & 15.42 & 22.88 & 28.44 & 37.62 & 45.38 & 53.10 & 60.50 & 66.32 & 75.56 \end{pmatrix}
$$

The coefficients of the 0th yarn of the second yarn group are as follows:

$$
M = \begin{bmatrix}
-0.174 & 2.29 \\
0.940 & 0.94 \\
0.655 & 8.14 \\
-0.531 & 13.03 \\
-0.882 & 10.08 \\
-0.137 & 1.98 \\
0.860 & 0.92 \\
0.739 & 7.56 \\
-0.502 & 13.03 \\
-0.846 & 10.11
\end{bmatrix}
\qquad
N = \begin{bmatrix}
-0.490 & 7.83 \\
-0.577 & 6.96 \\
0.704 & 0.38 \\
0.631 & 5.63 \\
0.607 & 9.14 \\
0.514 & 0.90 \\
0.760 & 3.05 \\
-0.716 & 8.92 \\
-0.546 & 3.62 \\
0.780 & 0.44
\end{bmatrix}
$$

$$
b = \begin{pmatrix} 0 & 7.76 & 15.42 & 22.88 & 28.44 & 37.62 & 45.38 & 53.10 & 60.50 & 66.32 & 75.56 \end{pmatrix}.
$$

The coefficients of other yarns in the same yarn group could be described with Equations 3.20 and 3.21.

3.2.4.1.2 Preform 6 × 6

For the 6 × 6 preform, the yarn group is six and there are eight yarns in each group. The coefficients of Equation 3.18 are as follows:

First group,

$$M = \begin{bmatrix} -0.967 & 11.13 \\ 0.274 & 1.11 \\ 0.897 & 2.79 \\ -0.132 & 11.94 \end{bmatrix} \quad N = \begin{bmatrix} 0.947 & 0.83 \\ 0.262 & 10.64 \\ -0.979 & 12.25 \\ -0.234 & 2.26 \end{bmatrix}$$

$$b = \begin{pmatrix} 0 & 10.36 & 16.50 & 26.70 & 32.82 \end{pmatrix}$$

Second group,

$$M = \begin{bmatrix} -0.622 & 3.93 \\ 0.865 & 0.56 \\ 0.558 & 10.02 \\ -0.820 & 13.02 \end{bmatrix} \quad N = \begin{bmatrix} 0.660 & 0.53 \\ -0.667 & 3.34 \\ -0.586 & 12.74 \\ 0.537 & 10.02 \end{bmatrix}$$

$$b = \begin{pmatrix} 0 & 5.42 & 16.36 & 21.74 & 32.82 \end{pmatrix}$$

Third group,

$$M = \begin{bmatrix} 0.698 & 0.53 \\ 0.791 & 5.57 \\ -0.669 & 12.83 \\ -0.811 & 8.01 \end{bmatrix} \quad N = \begin{bmatrix} 0.762 & 7.21 \\ -0.789 & 12.71 \\ -0.682 & 5.47 \\ 0.721 & 0.56 \end{bmatrix}$$

$$b = \begin{pmatrix} 0 & 7.22 & 16.40 & 23.6 & 32.82 \end{pmatrix}$$

Fourth group,

$$M = \begin{bmatrix} 0.884 & 0.53 \\ 0.622 & 9.56 \\ -0.737 & 12.83 \\ -0.646 & 3.90 \end{bmatrix} \quad N = \begin{bmatrix} -0.868 & 9.43 \\ 0.635 & 0.56 \\ 0.734 & 3.90 \\ -0.646 & 12.80 \end{bmatrix}$$

$$b = \begin{pmatrix} 0 & 10.22 & 15.48 & 27.60 & 32.82 \end{pmatrix}$$

Fifth group,

$$
M = \begin{bmatrix} -0.748 & 11.75 \\ -0.623 & 3.98 \\ 0.606 & 0.92 \\ 0.583 & 7.15 \end{bmatrix} \quad N = \begin{bmatrix} 0.660 & 4.08 \\ -0.667 & 11.26 \\ -0.586 & 7.23 \\ 0.537 & 0.83 \end{bmatrix}
$$

$$
b = \begin{pmatrix} 0 & 10.36 & 16.50 & 26.70 & 32.82 \end{pmatrix}
$$

Sixth group,

$$
M = \begin{bmatrix} 0.933 & 2.75 \\ -0.063 & 12.12 \\ -1.111 & 11.72 \\ 0.338 & 0.59 \end{bmatrix} \quad N = \begin{bmatrix} 0.986 & 1.27 \\ 0.247 & 11.17 \\ -1.046 & 12.74 \\ -0.155 & 2.26 \end{bmatrix}
$$

$$
b = \begin{pmatrix} 0 & 10.04 & 16.40 & 26.42 & 32.82 \end{pmatrix}
$$

3.2.4.2 Stress–Strain Relationship of Braiding Yarn

The experimental tensile curve of braiding yarn is shown in Figure 3.15. The fitted polynomial equation is as follows:

$$
y = 0.52 + 335.74x - 1973.80\,x^2 + 30697.98\,x^3 - 1.51 \times 10^5 x^4
$$

where y is load (in Newtons) and x is uniaxial tensile strain.

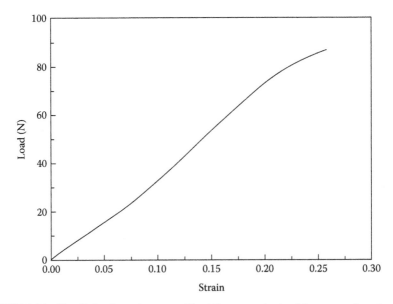

FIGURE 3.15   Tensile load–strain curve of braiding yarn obtained from experiment.

3.2.4.3 Predicted Results

When the strain of braiding yarn is increased from 0 to its failure strain, a series of loads, strain of braided preforms, and uniaxial tensile curves could be obtained. Comparisons between theoretical and experimental tensile curves for 6 × 4 preform and 6 × 6 preform are shown in Figures 3.16 and 3.17, respectively. It can be seen that the method described in this chapter could predict the tensile curve of 3D four-step 1 × 1 braided preform precisely in a simple and intuitive way.

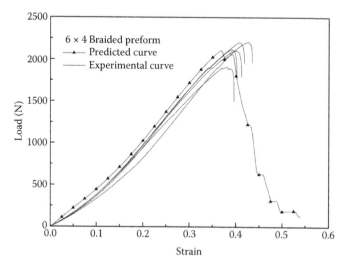

FIGURE 3.16 Comparison between predicted uniaxial tensile load–strain curve of braided preform and experimental ones (6 × 4 preform).

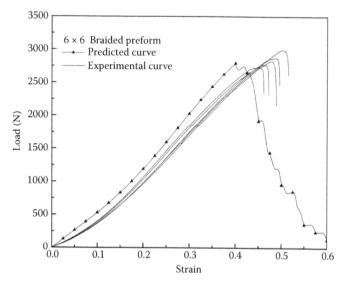

FIGURE 3.17 Comparison between predicted uniaxial tensile load–strain curve of braided preform and experimental ones (6 × 6 preform).

3.3 DESIGNING OF BRAIDED STRUCTURES AND EXAMPLES

In this section, we take the fatigue behaviors of 3D braided composite materials as an example for demonstrating the designing of braided structures.

The representative unit cells (RUCs) model has been developed based on the three-cell model proposed by Wang and Wang [10] to compute the three-point bending fatigue behaviors. A user-defined material (UMAT) subroutine that characterizes the stiffness matrix and fatigue damage evolution of the 3D braided composite was also developed. The resultant stress and strain distributions were analyzed to show the effect of stress and strain concentration on overall fatigue failure behavior. The stiffness degradation and displacement change were compared and analyzed under three stress levels. The results from this methodology were compared with experimental data from the literature for model verification.

3.3.1 NUMERICAL MODELS AND MATERIAL PROPERTIES OF REPRESENTATIVE UNIT CELLS

3.3.1.1 Modeling Strategy of Four-Step Three-Dimensional Braided Composite Structures

Based on results from the literature [4, 11–13], the four-step 3D braided composite can be divided into three representative regions, in which the braided structure is cycled during each braiding pitch (step 0 through step 4) along the braiding direction. Therefore, three types of RUCs, including interior RUC, surface RUC, and corner RUC, can be used to precisely analyze the mechanical performance during bending fatigue modeling, shown in Figure 3.18. The interior and exterior of braided structure are colored grey and green, respectively, and blue represents the yarn morphology in different RUCs.

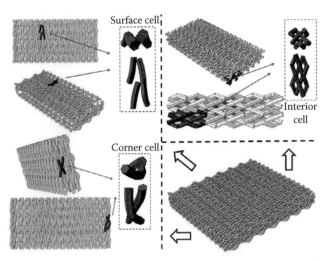

FIGURE 3.18 Representative unit cell distributions on the interior and exterior of braided structure.

The braided architecture and material property in this chapter were determined according to authors' previous experimental work [14]. Carbon fiber tows (Toray®, T300-6K) were used to braid the preform with the four-step 1 × 1 braiding procedure. The braided array was 29 × 5 (*m* = 29, *n* = 5). The braiding angle and pitch length of the preform were 28° ± 3° and 3.2 ± 0.2 mm, respectively. The composites were fabricated with epoxy resin (type: TDE-85) using resin transfer molding under ambient temperature. The curing process followed a stepwise program of 130°C for 2 hours, 150°C for 1 hour, 160°C for 8 hours, and 180°C for 3 hours. The measured fiber volume fraction was about 58%.

The following assumptions were made for the modeling: (1) the yarns are uniform along the length direction; (2) the braiding process is stable and all yarns are in the same jamming condition, ensuring the uniformity of braided structure; (3) all yarns in the braided composite have identical constituent material, size, and flexibility; and (4) the yarns are regarded as transversely isotropic, and the matrix is regarded as isotropic. The mechanical properties of fiber and matrix are listed in Table 3.2.

3.3.1.2 Structure Parameters and Volume Fraction of Representative Unit Cells

The yarn structures of interior, surface, and corner RUCs are depicted in Figure 3.19. The expanded length of preform in one braiding cycle is called braiding pitch. The three types of RUCs extend a braiding pitch along *X*, *Y*, and *Z* directions after a braiding cycle, forming the entire braiding cycle. Each yarn in the RUCs has a certain spatial orientation. By obtaining the angle, the mechanical performance of braiding structure can be calculated.

The yarns of interior RUCs are straight after jamming, as illustrated in Figure 3.19a. The interior braiding angles ±γ are defined by the angles between straight yarns and the Z axis. The angles ±φ are the projections of yarns on the *X–Y* plane and the *X* axis. Based on the study by Zhang, Xu, and Chen [11], the yarns of interior RUCs are parallel to ±45° cross section of the whole structure so that φ is 45°. In Figure 3.19b and c, the yarns on the surface and corner RUCs are spatial curves comprising a straight line and a segment of helix. However, the helix segment shows little influence on the entire braided structure, so it can be simplified as a straight line. The angles ±θ and ±β are the surface and corner braiding angles, respectively, and they are measured between the simplified straight yarns and the

---

TABLE 3.2

Mechanical Parameters of Fiber and Matrix

Property	Fiber (T300-6K)	Matrix (TDE-85)
Longitudinal modulus (GPa)	230	4.55
Transverse modulus (GPa)	13.8	4.55
Plastic strain (%)	—	2.3
Shear modulus (GPa)	9	1.75

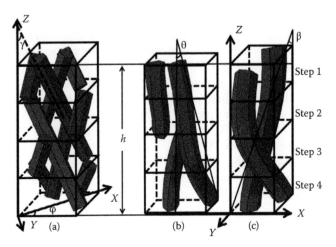

FIGURE 3.19 Yarn morphology in (a) interior representative unit cell (RUC), (b) surface RUC, and (c) corner RUC.

Z axis. The projected angles, φ, are $0°$ and $90°$. All the angles are related to braiding angle α, as shown in Equation 3.34:

$$\tan\alpha = \frac{\sqrt{2}}{2} \cdot \tan\gamma = \frac{6\sqrt{2}}{\pi} \cdot \tan\theta = \frac{6\sqrt{2}}{\pi} \cdot \tan\beta \qquad (3.34)$$

As shown in Figure 3.20, the interior, surface, and corner RUCs have different volume fractions during braiding. The volume fraction can be calculated from Equations 3.35 through 3.37:

$$V_I = \frac{2 \cdot (m-1) \cdot (n-1)}{2 \cdot m \cdot n + m + n} \qquad (3.35)$$

$$V_S = \frac{3 \cdot (m+n-2)}{2 \cdot m \cdot n + m + n} \qquad (3.36)$$

$$V_C = \frac{4}{2 \cdot m \cdot n + m + n} \qquad (3.37)$$

V_I, V_S, and V_C are volume fractions of the interior, surface, and corner RUCs with respect to the whole structure, respectively. It must be noted that m and n are the numbers of columns and rows, respectively, in the cross section of braided perform ($m = 29$, $n = 5$). The volume fractions of interior, surface, and corner RUCs were calculated to be 69.14%, 29.63%, and 1.23%, respectively.

3.3.1.3 Mechanical Properties of Representative Unit Cells in Global Coordinate System

From the earlier description, yarns keep certain spatial orientations in RUCs. According to the fourth assumption, axis-1 is defined along the yarn path line, and

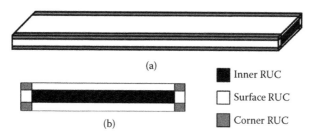

(a)

■ Inner RUC

□ Surface RUC

▨ Corner RUC

(b)

FIGURE 3.20 Interior, surface, and corner representative unit cell distributions in composite: (a) oblique view and (b) cross-sectional view.

axis-2 and axis-3 are perpendicular to the axis-1 direction. To obtain the material property of the entire braided structure, it is inadequate to determine the material property of yarns and matrix in local coordinates. The transpose of matrix T_{ij} is applied to calculate the material property in global coordinate system where axis X, axis Y, and axis Z are along the width, height, and length directions of braided fabric, respectively. Figure 3.21 depicts the relations between local and global coordinates.

The compliance matrix in the local coordinate system is transformed into the global coordinate system through the following equations (Equations 3.38 and 3.39):

$$[S]_g = [T_{ij}]_S [S]_1 [T_{ij}]_S^T \tag{3.38}$$

where $[S]_1$ and $[S]_g$ are the compliance matrices in local and global coordinate systems, respectively. $[T_{ij}]_S$ is the transformation matrix, and it is given by

$$[T_{ij}]_S = \begin{bmatrix} l_1^2 & l_2^2 & l_3^2 & l_2 l_3 & l_3 l_1 & l_1 l_2 \\ m_1^2 & m_2^2 & m_3^2 & m_2 m_3 & m_3 m_1 & m_1 m_2 \\ n_1^2 & n_2^2 & n_3^2 & n_2 n_3 & n_3 n_1 & n_1 n_2 \\ 2m_1 n_1 & 2m_2 n_2 & 2m_3 n_3 & m_2 n_3 + m_3 n_2 & n_3 m_1 + n_1 m_3 & m_1 n_2 + m_2 n_1 \\ 2n_1 l_1 & 2n_2 l_2 & 2n_3 l_3 & l_2 n_3 + l_3 n_2 & n_3 l_1 + n_1 l_3 & l_1 n_2 + l_2 n_1 \\ 2l_1 m_1 & 2l_2 m_2 & 2l_3 m_3 & l_2 m_3 + l_3 m_2 & l_1 m_3 + l_3 m_1 & l_1 m_2 + l_2 m_1 \end{bmatrix} \tag{3.39}$$

in which $l_1 = \cos(\eta_X) \cdot \sin(\eta_Z)$, $m_1 = \sin(\eta_X) \cdot \sin(\eta_Z)$, $n_1 = \cos(\eta_Z)$, $l_2 = -\sin(\eta_X)$, $m_2 = \cos(\eta_X)$, $n_2 = 0$, $l_3 = -\cos(\eta_X) \cdot \cos(\eta_Z)$, $m_3 = -\sin(\eta_X) \cdot \cos(\eta_Z)$, and $n_3 = \sin(\eta_Z)$. The angles η_X and η_Z for each type of RUC are different, and the angle values are listed in Table 3.3.

The material property of each type of RUCs can be calculated using the aforementioned method, and the stiffness matrix is computable by the rule of mixture. The total stiffness of 4S3DBC (4-step 3-dimenaional braided composite) was calculated by the volume averaging method, shown in Equation 3.40:

$$[C]_{\text{total}} = \Sigma V_n [C_n]_g \tag{3.40}$$

where n indicates the different kinds of unidirectional composites in interior, surface, and corner RUCs. V_n is the volume proportion of RUCs. The stiffness matrix $[C]$ was calculated using $[C] = [S]^{-1}$.

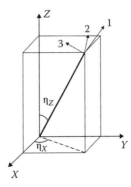

FIGURE 3.21   Relationship between local and global coordinates.

TABLE 3.3
Yarn Angles of Interior, Surface, and Corner RUCs

RUC	η_x	η_y
Interior	$\pm\phi$	$\pm\gamma$
Surface	$0°, 90°$	$\pm\theta$
Corner	$0°, 90°$	$\pm\beta$

3.3.2   FINITE ELEMENT ANALYSES

3.3.2.1   Modeling Details

Finite element modeling, calculation, and analysis of braided composite materials were conducted on the commercially available finite element software package ABAQUS/Standard (version 6.11), and the operating system platform was LINUX. A UMAT for ABAQUS/Standard was compiled with the Intel Visual FORTRAN Compiler Professional 11.1.065. The user-defined subroutine UMAT was used to define the constitutive relationship and damage accumulation and failure criterion (RUCs) of the four-step 3D braided composite under different loading conditions. The flowchart of the accumulated fatigue damage that was used for developing a computer code to calculate needed parameters is shown in Figure 3.22.

3.3.2.2   Load and Mesh

The loading pattern was designed following authors' previous experimental work [15]. The three-point bending fatigue tests were performed under a sinusoidal cyclic loading with a stress ratio (ratio of the minimum stress to the maximum stress in one cycle) of 0.1 and a frequency of 3 Hz. The bending elastic modulus and failure strength were measured as 97.6 GPa and 615 MPa, respectively, in the experiment. Stress level is described as the ratio of the maximum applied stress in one cycle to the maximum static bending stress. Three stress levels, 60%, 70%, and 80%, were chosen for bending fatigue modeling. Lower stress levels were tried, but they were not

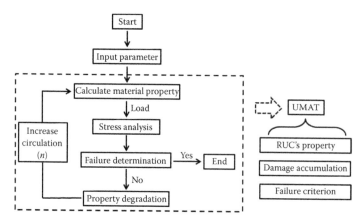

FIGURE 3.22 Flowchart representing the general strategy for damage accumulation modeling.

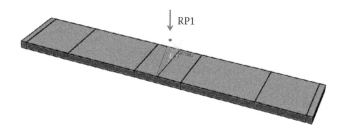

FIGURE 3.23 Geometrical model of composite under three-point bending cyclic loading.

chosen because the specimens did not fail after the bending fatigue loading cycles of 10^6.

The geometrical model was generated in the commercial software ABAQUS/CAE. It can be used to create the model and settle the finite element details such as contact property, boundary condition, mesh scheme, and strain increment. For linear contact in the three-point bending test, a simplified loading model was adopted to avoid the complex contact calculation during finite element analysis (FEA) and save computational time. Reference point (RP1) was created in the center of the upper surface of the finite element model, as shown in Figure 3.23. RP1 was coupled with the center line of the upper surface of the model by the "structural distributing" function in ABAQUS. The load was a concentrated force applied at the RP1 point. The loading pattern was following the experimental procedure. The lower surface of the composite model was set as the boundary condition of "Displacement/Rotation."

The eight-node reduced integration element (C3D8R) was applied for mesh discretization. In mesh discretization, the calculation accuracy was not merely determined by mesh size but influenced by the size of RUC. For example, when mesh size was much smaller than the size of RUC, the accuracy was low. Therefore, local segmentation was adopted for the composite plate with the fine segmentation in the stress concentrated areas. The fine mesh size was $0.5 \times 0.5 \times 1$ mm, and the rest part

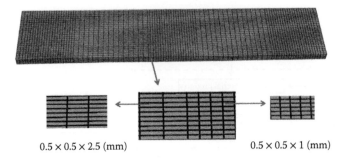

0.5 × 0.5 × 2.5 (mm)   0.5 × 0.5 × 1 (mm)

FIGURE 3.24 Mesh scheme adopted in modeling, and fine mesh sizes are applied in the stress concentrated areas.

was 0.5 × 0.5 × 2.5 mm. This mesh scheme can ensure the calculation accuracy as well as reduce the computational cost. As shown in Figure 3.24, the element numbers of the plate and rollers were 28,200.

3.3.2.3   Damage Accumulation and Failure Criterion

The damage growth in composites has been regarded as a metric of fatigue damage accumulation, because it is an easy and nondestructive measurement that reflects a variety of complicated damage mechanisms evolving in the microstructural level [16,17]. The expression of damage growth under fatigue loading was improved by Paepegem and Degrieck [18], as given in Equation 3.41.

$$\frac{\mathrm{d}D}{\mathrm{d}n} = \begin{vmatrix} \dfrac{A \cdot \left(\dfrac{\Delta\sigma}{\sigma_{TS}}\right)^{c}}{(1-D)^{b}} & \text{in tension} \\ 0 & \text{in compression} \end{vmatrix} \qquad (3.41)$$

where D is the local damage variable; n is the number of cycles; $\Delta\sigma$ is the amplitude of the applied cyclic loading; σ_{TS} is the tensile strength; and A, b, and c are three material constants determined by least-square fitting.

To obtain the relation between D and n, we can integrate both sides of the equation. The final expression of the local damage variable is the following equation:

$$D(n) = 1 - [1 - (1+b)A(\frac{\Delta\sigma}{\sigma_{TS}})^{c} n]^{\frac{1}{1+b}} \qquad (3.42)$$

The fatigue failure criterion for combined cyclic stress that Hashin [19] proposed based on the static failure criterion is applicable in justifying different failure modes. Shokrieh and Lessard [20] extended Hashin's work by adding the effects of material nonlinearity on the fatigue failure criterion. Therefore, it is chosen to define the failure of 4S3DBC under three-point bending. Based on the bending mechanism, the bending motion can be simplified as a combination of tension and compression. Equations 3.43 through 3.46 are the fatigue failure criterions of fiber and matrix:

1. When fiber tensile fatigue failure occurs, $\sigma > 0$,

$$\left(\frac{\sigma}{F_T(n,\sigma,R)}\right)^2 + \left(\frac{\dfrac{\sigma_{XY}^2}{2E_{XY}(n,\sigma,R)}}{\dfrac{S_{XY}^2(n,\sigma,R)}{2E_{XY}(n,\sigma,R)}}\right)^2 + \left(\frac{\dfrac{\sigma_{XZ}^2}{2E_{XZ}(n,\sigma,R)}}{\dfrac{S_{XZ}^2(n,\sigma,R)}{2E_{XZ}(n,\sigma,R)}}\right)^2 \geq 1 \qquad (3.43)$$

2. When fiber compressive fatigue failure occurs, $\sigma < 0$,

$$\left(\frac{\sigma}{F_C(n,\sigma,R)}\right)^2 \geq 1 \qquad (3.44)$$

3. When matrix tensile fatigue failure occurs, $\sigma > 0$,

$$\left(\frac{\sigma}{M_T(n,\sigma,R)}\right)^2 + \left(\frac{\dfrac{\sigma_{XY}^2}{2E_{XY}(n,\sigma,R)} + \dfrac{3}{4}\alpha\sigma_{XY}^4}{\dfrac{S_{XY}^2(n,\sigma,R)}{2E_{XY}(n,\sigma,R)} + \dfrac{3}{4}\alpha S_{XY}^4(n,\sigma,R)}\right)^2 + \left(\frac{\sigma_{YZ}}{S_{YZ}(n,\sigma,R)}\right)^2 \geq 1 \quad (3.45)$$

4. When matrix compressive fatigue failure occurs, $\sigma < 0$,

$$\left(\frac{\sigma}{M_C(n,\sigma,R)}\right)^2 + \left(\frac{\dfrac{\sigma_{XY}^2}{2E_{XY}(n,\sigma,R)} + \dfrac{3}{4}\alpha\sigma_{XY}^4}{\dfrac{S_{XY}^2(n,\sigma,R)}{2E_{XY}(n,\sigma,R)} + \dfrac{3}{4}\alpha S_{XY}^4(n,\sigma,R)}\right)^2 + \left(\frac{\sigma_{YZ}}{S_{YZ}(n,\sigma,R)}\right)^2 \geq 1 \quad (3.46)$$

where σ is the stress during the nth loading cycle; the subscripts X, Y, and Z are the directions of global coordinate; $F_k(n,\sigma,R)$ and $M_k(n,\sigma,R)$ are the residual fatigue strengths of fiber and matrix, respectively, under fatigue loading; k represents tension and compression when $k = T$ and $k = C$, respectively; $E(n,\sigma,R)$ and $S(n,\sigma,R)$ are the normal and shear residual fatigue stiffness; and α is the nonlinearity parameter, which is 3.7 mPa^{-3}[21] for epoxy resin. The strength constants of fiber and matrix have been obtained by experiment [22] and are listed in Table 3.4.

3.3.3 RESULTS AND DISCUSSIONS

After the earlier calculations (in Section 3.3.2), the maximum loading cycles are 101,500; 50,370; and 12,800 under 60%, 70%, and 80% stress levels, respectively.

TABLE 3.4
Mechanical Parameters of Fiber and Matrix

Property		Fiber	Matrix
Longitudinal strength (MPa)	Tension	3530	80
	Compression	−5300	−370
Transverse strength (MPa)	Tension	1000	80
	Compression	−2000	−370
Shear strength (MPa)	Longitudinal	1000	100
	Transverse	800	100

3.3.3.1   Stress Distribution and Deflection Change

Figure 3.25 presents the stress distribution of a composite panel at three stress levels (60%, 70%, and 80%). Five specific cycles (first cycle and the cycles of 30%, 60%, 90%, and 97% life span) were chosen to observe the stress change during the life span. Because the stress control loading pattern was chosen, the loading stress of specimen was kept constant with the increase of the loading cycle. The stress change of the composite panel is mild with the fatigue damage accumulation and the decrease of flexural modulus for different cycles. As shown in Figure 3.25, stress concentrates on the middle of the specimen and on both ends of the bottom areas where the constraint was applied. The stress scale bar reflects the relations of stress value and color. Under different stress levels, the maximum stress increased from 310 MPa for 60% stress level to 390 MPa for 80% stress level.

Figure 3.26 is the bending strain distribution of the composite panel under three stress levels. The same specific cycles as stress analysis were chosen. The strain scale bar was set from 0.002% to 2% to compare the strain change under the same strain level. It can be observed that the areas with larger strain are inconsistent with the stress distribution discussed earlier, that is, the middle of the specimen and both ends of the bottom areas. The strain value goes up with the cycles. For different stress levels, the strain distributions at a cycle of 97% life span show the most differences, and the strain increases sharply at the end of the life span.

3.3.3.2   Bending Displacement

Comparing the bending displacement and loading cycle under 60%, 70%, and 80% stress levels in Figure 3.27, the trend is similar, although the maximum displacement and loading cycle vary. According to the fatigue damage mechanism, three stages are defined to describe the fatigue process. In stage I, initial displacement occurs; then, the curves reach a plateau with few displacement changes in stage II. In the final stage (stage III), the displacement increases sharply, indicating that the material is stepping into the last period of the life span. The difference of the three curves can be used to analyze the lifetime of the composite panels under different stress levels. In Figure 3.27c, stage III starts at the 60,991st cycle, which is 60% of the life span ($N = 101,500$), whereas for 70% and 80% stress levels stage III is initiated at the 40,296th and 11,550th cycles, which are 80% and 90% of the life span, respectively. It can be clearly observed that the curves slow down with the decrease of stress

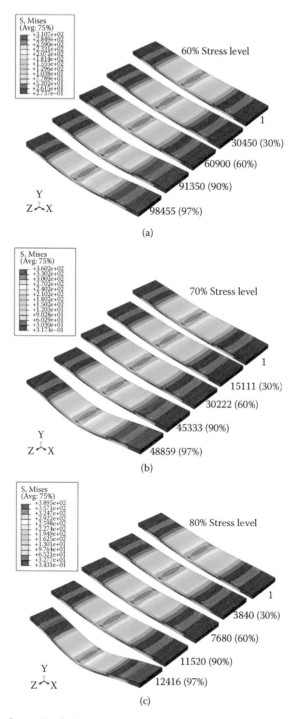

FIGURE 3.25 Stress distribution of specimen of different stress levels: (a) 60% stress level, (b) 70% stress level, and (c) 80% stress level.

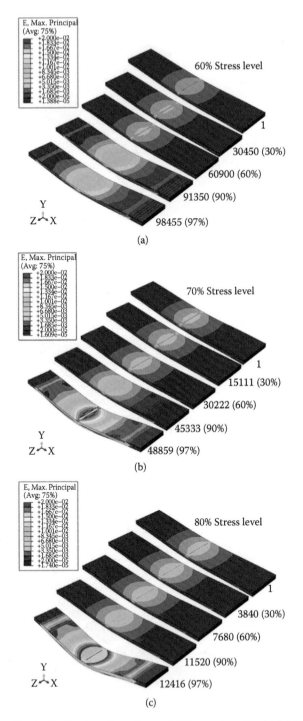

FIGURE 3.26 Strain distribution of specimen of different stress levels: (a) 60% stress level, (b) 70% stress level, and (c) 80% stress level.

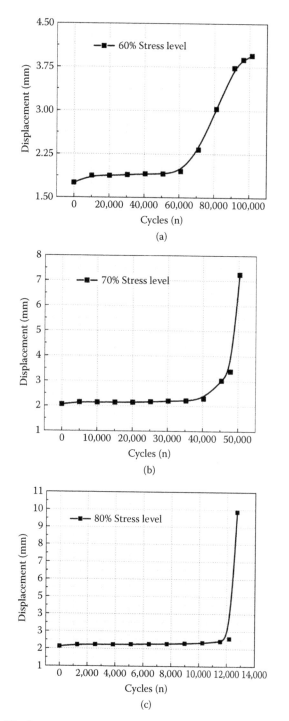

FIGURE 3.27 Displacement versus loading cycle relations under (a) 60%, (b) 70%, and (c) 80% stress levels.

levels. During bending fatigue loading, with a higher stress level, the material has faster damage accumulation. Besides, the material presents a sudden failure under higher stress levels. The higher stress level leads to a larger displacement, resulting in fatigue failure of the entire composite.

3.3.3.3 Stiffness Degradation

Figure 3.28 shows the comparison between experimental and finite element method (FEM) results. It is found that there is a good agreement between experimental results and FEM, which indicates that the usage of UMAT is reliable in modeling damage accumulation and fatigue failure. Similar to Figure 3.27, the stiffness degradation curve can be divided into three stages. In stages I and II, the bending stiffness of the specimens decreased with the increase in loading cycles. According to fatigue tests, the damage of composite includes fiber tows breakages, debonding at the interface between fiber tows and matrix, and matrix failure. All of the damage modes might cause stiffness degradation of the material. The middle stage involves a slow decrease in stiffness during a long time. In the final stages, there exists a sharp decrease in stiffness, which corresponds to a short time. Taking the fatigue damage accumulation of the braided composite into consideration, in the initial stage, the resin crack appears after several testing cycles, which leads to the decline in stiffness. With the loading of the cyclic test, the surface layer of the resin crack extends to the interface due to the stress concentration in the intermediate stage. This leads to resin–yarn delamination and partial destruction of the original yarn, and this stage takes a longer time. In the final stage, the cracks continue to grow along the resin–yarn interface due to the effects of stress concentration. Under further testing, the progressive damage of yarns occurs until ultimate breakage. Then, the braided composite no longer bears the cyclic loading and this is followed by complete failure.

3.3.3.4 Bending Modulus

Specific cycles were chosen to observe the relation between load and displacement under 60%, 70%, and 80% stress levels, that is, the first cycle and the cycles of 10%, 20%, 30%, 40%, 50%, 60%, 70%, 80%, 90%, 95%, and 97% life span, seen in Figure 3.29. The slope of each line represents the bending stiffness at specific cycles. From these lines, the changes of displacement and stiffness can be observed under three stress levels. At 60% stress level, the lines are densely distributed in two regions, that is, before the 60,900th cycle (60% of life span) and after the 91,350th cycle (90% of life span). However, there is only one densely distributed region for 70% and 80% stress levels, and the largest displacement and slope change appear at the last cycles, which are the 48,859th and 12,416th cycles for 70% and 80% stress levels, respectively. It illustrates that at higher stress levels (70% and 80% stress levels) fatigue damage is gradually accumulated until the deformation and stiffness degradation dramatically change at final failure. However, for lower stress levels (60% stress level), the loading is insufficient to cause failure after the large deformation and stiffness degradation, so the damage is continually accumulated. Therefore, the second densely distributed region is formed where the deformation and stiffness degradations are subjected to a steady variation.

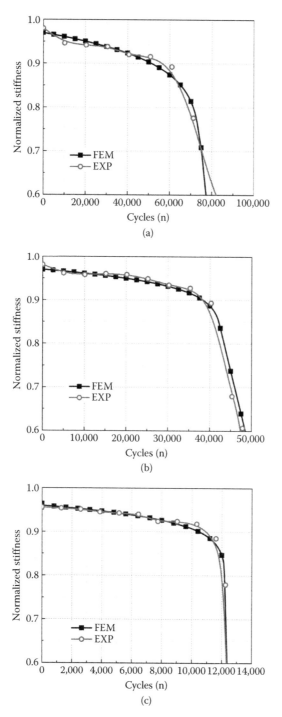

FIGURE 3.28 Experimental and finite element method results of stiffness degradation curve under (a) 60%, (b) 70%, and (c) 80% stress levels.

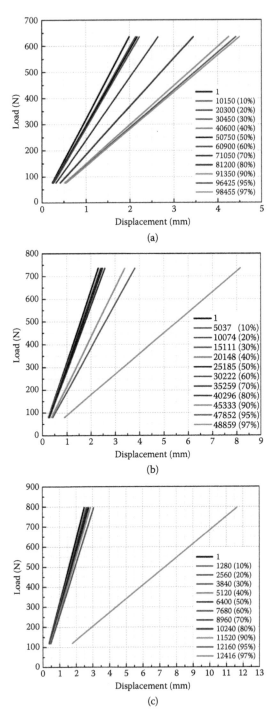

FIGURE 3.29 Load–displacement curves at specific cycles under (a) 60%, (b) 70%, and (c) 80% stress levels.

3.3.4   SOME SUGGESTIONS

This section presents an example for demonstrating the design of 3D braided composite materials by building three types of RUC models based on the microstructure of braided composite materials. A UMAT was written and incorporated with commercial finite element code ABAQUS/Standard to calculate the stress and strain distributions, bending displacement change, stiffness degradation, and load–displacement curves at three different stress levels. Stress concentration and large deflection were found to occur at the middle of the composite panel and both ends of the bottom areas. At higher stress levels, the entire composite was subjected to larger deformation and faster damage accumulation, resulting in prompt fatigue failure. Good agreements of stiffness degradation curves were reached between the finite element simulation and experiment. It is inferred that it is possible to extend the RUC model to the bending fatigue performance design of other 3D braided composite structures. For more information, the studies by Wu and Gu [23] and Wu et al. [24] are suggested for further reading.

3.4   SOFTWARE TOOLS

As mentioned in Section 3.3, FEA software is often employed to design and analyze the microstructures and mechanical behaviors of braided composite materials and structures.

Usually, the preprocessor software of FEA is called computer-aided engineering (CAE) software. For example, the CAE software of CATIA is often used to connect with the commercial FEA software of ABAQUS to analyze the mechanical responses of 3D braided composite materials.

Now, the software of WiseTex suit developed by Dr. Stepan Lomov from the University of Leuven, Belgium, is also often used to construct the geometrical model of 3D braided composite materials.

Here, we present an example of establishing the geometrical model of 3D braided composite materials developed with the CAE software of Pro/E at the mesostructure level for applications in ballistic impact damage analyses [25].

In 3D rectangular braided composite materials, the spatial configuration of the braiding yarn of each yarn group is visualized in Figure 3.30. The complete microstructure geometrical model of 3D braided preform is shown in Figure 3.31. The microstructure geometrical model of 3D braided composites is shown in Figure 3.32, in which the yarns are transparent.

The mesh scheme for the 12 × 4 braided composites is shown in Figure 3.33a and that for the 12 × 4 braided preform is shown in Figure 3.33b. There are about 2,500,000 tetrahedral solid elements in the microstructure geometrical model. Based on the assumption of a perfect bonding between yarn and matrix, the elements of yarns and matrix share the same nodes and surface in meshing.

Figure 3.34 is the ballistic perforation process and damage evolution of 12 × 4 braided preform and composites in distal side at a strike velocity of 391 m/s (the left side shows the composite, and the right side shows the preform in the composite). Figure 3.35 is for 12 × 6 type 3D composites at a strike velocity of 390 m/s (the left

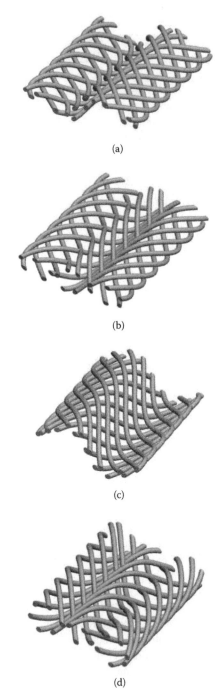

FIGURE 3.30 Visualization of spatial configuration of the braiding yarn of each yarn group for a 12 × 4 braided preform: (a) visualization of group 1, (b) visualization of group 2, (c) visualization of group 3, and (d) visualization of group 4.

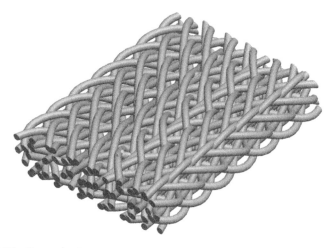

FIGURE 3.31 Integral microstructure geometrical model of 12 × 4 braided preform.

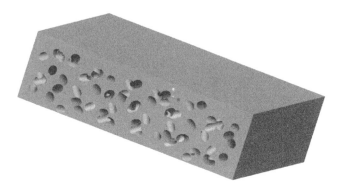

FIGURE 3.32 Microstructure of 12 × 4 type three-dimensional (3D) braided composite (yarns are transparent).

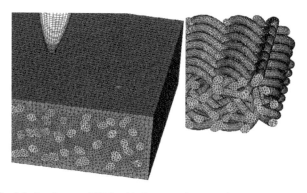

FIGURE 3.33 Mesh scheme of 3D braided composite materials.

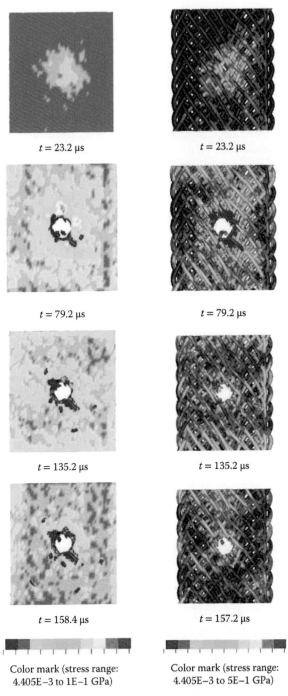

$t = 23.2\ \mu s$ $t = 23.2\ \mu s$

$t = 79.2\ \mu s$ $t = 79.2\ \mu s$

$t = 135.2\ \mu s$ $t = 135.2\ \mu s$

$t = 158.4\ \mu s$ $t = 157.2\ \mu s$

Color mark (stress range: Color mark (stress range:
4.405E−3 to 1E−1 GPa)   4.405E−3 to 5E−1 GPa)

FIGURE 3.34   Ballistic perforation process of 12 × 4 braided composites and their reinforcing preform in distal side at a strike velocity of 391 m/s (left side: 3D braided composite; right side: 3D braided preform in the composite).

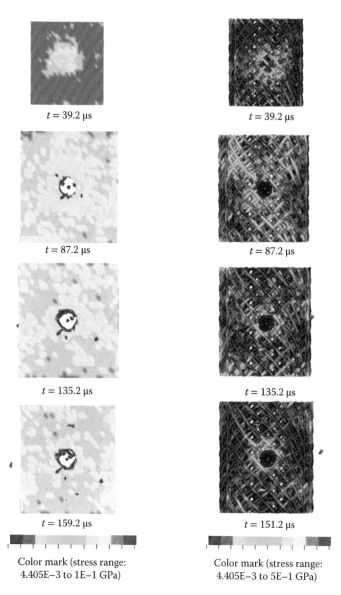

$t = 39.2\ \mu s$ $t = 39.2\ \mu s$

$t = 87.2\ \mu s$ $t = 87.2\ \mu s$

$t = 135.2\ \mu s$ $t = 135.2\ \mu s$

$t = 159.2\ \mu s$ $t = 151.2\ \mu s$

Color mark (stress range: Color mark (stress range:
4.405E–3 to 1E–1 GPa)   4.405E–3 to 5E–1 GPa)

FIGURE 3.35 Ballistic perforation process of 12×6 braided composites and their reinforcing preform in distal side at a strike velocity of 390 m/s (left side: 3D braided composite; right side: 3D braided preform in the composite).

side shows the composite, and the right side shows the preform in the composite). From the FEA simulated results, the damage modes of fiber breakage, fiber pullout, and matrix crack could be observed and compared with those in Figure 3.36. The damage pattern in Figure 3.36 is the local magnification near the perforating hole.

From the comparison of results obtained from the microstructure model and other models based on continuum assumption, the microstructure model is seen to be more

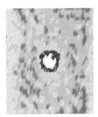
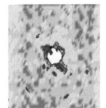

(a) Striking side   (b) Distal side

(1) Simulated damage morphology of inclined lamina of 12 × 4 braided composites

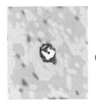
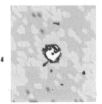

(a) Striking side   (b) Distal side

(2) Simulated damage morphology of inclined lamina of 12 × 6 braided composite

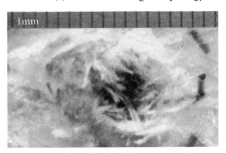

Damage morphology near penetration hole in   Damage morphology near penetration hole
striking side   in distal side

(3) Local magnification of damage pattern around the perforating hole

FIGURE 3.36   Simulated and observed damage morphologies of 3D braided composite (strike velocity: 12 × 4 braided composite, 391 m/s; 12 × 6 braided composite, 390 m/s).

precise in the calculation of energy absorption, damage propagation, and deformation development. This model can be extended to the simulation of debris impact in spacecraft design because 3D braided composites have been widely used in supersonic aircraft and spaceships. Furthermore, the microstructure model can be utilized to simulate ballistic impact damage and dynamic responses of 3D braided composite structures (such as a beam, a cylinder, or a tube).

3.5   SUMMARY AND CONCLUSIONS

Three-dimensional braided preforms are one type of important reinforced textile structures for composite material manufacture. There are two types of 3D braided preforms: two-step and four-step 3D braided preforms. In this chapter, the stiffness matrix and tensile strength of braided preforms have been modeled and predicted from the microstructure model. The structure modeling strategies of the braided preforms have been presented. This chapter places more emphasis on the dynamic mechanical behaviors of braided composite materials. For example, the fatigue behaviors of braided composite materials have been numerically characterized from unit cell and full-size microstructure levels. The stiffness degradation and deformation evolution have been calculated from different-scale models. Some software tools for establishing the microstructure models of braided preforms and composite materials have also been presented.

ACKNOWLEDGMENTS

The authors acknowledge the financial supports from the National Science Foundation of China (grant number 11272087). The financial supports from the Foundation for the Author of National Excellent Doctoral Dissertation of PR China (number 201056), Shanghai Rising-Star Program (11QH1400100), and Fundamental Research Funds for the Central Universities of China are also gratefully acknowledged.

REFERENCES

1. Joon-Hyung Byun, Tsu-Wei Chou. Mechanics of textile composites. In: *Comprehensive Composite Materials* (editors-in-chief: Anthony Kelly, Carl Zweben), Volume 1: Fiber reinforcements and general theory of composites, Elsevier Ltd., Amsterdam, the Netherlands, 2000, 719–761.
2. Tsu-Wei Chou. *Microstructural Design of Fiber Composites*. Cambridge University Press, Cambridge, United Kingdom, 1992, 296–297.
3. Du GW, Chou TW, Popper P. Analysis of three-dimensional textile preforms for multidirectional reinforcement of composites. *Journal of Materials Science*, 1991, 26(13): 3438–3448.
4. Chen Li, Tao Xiaoming, Choy Choyloong. A structural analysis of three-dimensional braids. *Journal of China Textile University (Eng. Ed.)*, 1997, 14(3): 8–13.
5. Chen L, Li JL, Li XM, Qiu GX. Yarn architecture of 3-D rectangular braided preforms. *Acta Materiae Compositae Sinica*, 2000, 17(3): 1–5 (in Chinese).
6. Tao XM, Xian XJ, Ko FF. *Textile Structural Composites*. Science Press, Beijing, China, 2001, 292–304 (in Chinese).
7. Wang Y, Sun X. Digital-element simulation of textile processes. *Composites Science and Technology*, 2001, 61: 311–319.
8. Gu BH. Prediction of the uniaxial tensile curve of 4-step 3-dimensional braided preform. *Composite Structures*, 2004, 64(2): 235–241.
9. Li W, El Shiekh A. The effect of processes and processing parameters on 3-D braided preforms for composites. *SAMPE Quarterly*, 1988, 19: 22–28.
10. Wang Y, Wang ASD. On the topological yarn structure of 3-D rectangular and tubular braided preforms. *Composites Science and Technology*, 1994, 51(4): 575–586.
11. Zhang C, Xu X, Chen K. Application of three unit-cells models on mechanical analysis of 3D five-directional and full five-directional braided composites. *Applied Composite Materials*, 2013, 20(5): 803–825.
12. Li DS, Li JL, Chen L, Lu ZX, Fang DN. Finite element analysis of mechanical properties of 3D four-directional rectangular braided composites part 1: microgeometry and 3D finite element model. *Applied Composite Materials*, 2010, 17(4): 373–387.

13. Shokrieh MM, Mazloomi MS. A new analytical model for calculation of stiffness of three-dimensional four-directional braided composites. *Composite Structures*, 2012, 94(3): 1005–1015.

14. Zhao Q, Jin LM, Jiang LL, Zhang Y, Sun BZ, Gu BH. Bending fatigue behavior of four-step 3-D braided rectangular composite under different stress levels. *Journal of Reinforced Plastics and Composites*, 2011, 30(18): 1571–1582.

15. Sun BZ, Liu RQ, Gu BH. Numerical simulation of three-point bending fatigue of four-step 3-D braided rectangular composite under different stress levels from unit-cell approach. *Computational Materials Science*, 2012, 65: 239–246.

16. Lee L, Yang J, Sheu D. Prediction of fatigue life for matrix-dominated composite laminates. *Composites Science and Technology*, 1993, 46(1): 21–28.

17. Hwang W, Han KS. Fatigue of composites—fatigue modulus concept and life prediction. *Journal of Composite Materials*, 1986, 20(2): 154–165.

18. Van Paepegem W, Degrieck J. Fatigue degradation modelling of plain woven glass/epoxy composites. *Composites Part A: Applied Science and Manufacturing*, 2001, 32(10): 1433–1441.

19. Hashin Z. Fatigue failure criteria for combined cyclic stress. *International Journal of Fracture*, 1981, 17(2): 101–109.

20. Shokrieh MM, Lessard LB. Progressive fatigue damage modeling of composite materials, part I: modeling. *Journal of Composite Materials*, 2000, 34(13): 1056–1080.

21. Morancho JM, Salla JM. Relaxation in a neat epoxy resin and in the same resin modified with a carboxyl-terminated copolymer. *Journal of Non-Crystalline Solids*, 1998, 235–237(0): 596–599.

22. Zhang Y, Jiang LL, Sun BZ, Gu BH. Transverse impact behaviors of four-step 3-D rectangular braided composites from unit-cell approach. *Journal of Reinforced Plastics and Composites*, 2012, 31(4): 233–246.

23. Wu LW, Gu BH. Fatigue behaviors of four-step three-dimensional braided composite material: A meso-scale approach computation. *Textile Research Journal*, 2014, 84(18): 1915–1930.

24. Wu LW, Zhang F, Sun BZ, Gu BH. Finite element analyses on three-point low-cyclic bending fatigue of 3-D braided composite materials at microstructure level. *International Journal of Mechanical Sciences*, 2014, 84(July): 41–53.

25. Gu B. A microstructure model for finite element simulation of 3-D 4-step rectangular braided composite under ballistic penetration. *Philosophical Magazine*, 2007, 87(30): 4643–4669.

4 Braided Composites
Production, Properties, and Latest Developments

Sohel Rana and Raul Fangueiro

CONTENTS

ABSTRACT

This chapter presents an overview of braided composites. Different types of braided composites and their advantages and applications are briefly discussed. Production routes of both thermosetting and thermoplastic braided composites are presented, and various properties such as in-plane mechanical properties, shear performance, crashworthiness, frictional and wear behavior, and surface adhesion with other matrices are discussed. Developments on self-sensing braided composites are also included. The last section of the chapter presents multiscale braided composites, which have been developed recently. Summary and future directions of research in the field of braided composites are also included.

4.1   INTRODUCTION

Recently, composite materials made of braided textile preforms are attracting tremendous attention. The use of braided composites is highly advantageous in applications requiring high shear and torsional strength and stiffness (Cagri and Carey 2008). Other important advantages offered by braided composites are increased transverse strength and modulus, high damage tolerance and fatigue life, notch insensitivity, high fracture toughness, and possibility to develop complex and near net shape composite preforms (Mouritz et al. 1999).

Braided preforms have huge flexibility in terms of producing different shapes such as hollow tubular, stuffed tubular, flat, solid square, and irregularly shaped or fashioned solids, and many complex profiles can be produced such as I beams, H beams, Δ beams, channel beams, angle beams, ribbed and solid columns, tubes, plates, and so on (Zsigmond 2005). These various profiles can be produced by tube braiding with the help of a subsequent deforming process.

Braided composites are produced from textile preforms using braiding techniques. Different types of braided architectures that are used to produce braided composites are shown in Figure 4.1 (Zsigmond 2005). Braided composites produced from these various preforms are named accordingly.

Braided preforms can be produced with a high degree of automation involving computer-controlled and robotic mechanisms, to produce various complex shapes. The automation in braiding technology has significantly cut down the associated labor costs. Therefore, the overall production cost of braided composites has gone down due to automation in the preform production and manufacturing of near net shapes, which eliminate the additional costs incurred in tooling, machining, and so on.

Among the different braiding techniques, three-dimensional (3D) braiding is more suitable for production of complex and near net shapes. However, the production rate of 3D braiding is lower as compared to the two-dimensional (2D) braiding process and also the production cost is higher in the case of the former technique as

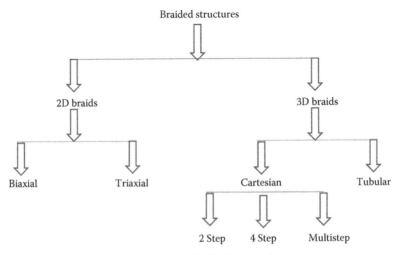

FIGURE 4.1   Types of fibrous preforms for braided composites.

each type of 3D braided preform needs different design and arrangement of the track configuration of the braiding machine. Therefore, some efforts have been recently made to produce complex shaped structural components using the conventional 2D braiding process in combination with the deforming process (Uozumi et al. 2005).

Regarding the mechanical performance of 2D and 3D braided preforms, the latter provides better delamination resistance, impact damage tolerance, and lower notch sensitivity due to the presence of through-the-thickness reinforcement (Mouritz et al. 1999). However, the in-plane mechanical properties of 3D braided composites are usually lower than those of 2D ones with equivalent in-plane fiber weight fraction, as a result of off-axis and highly crimped fiber configuration. Multiaxis 3D braided preforms have been developed recently incorporating ± bias yarn layers, to improve the in-plane mechanical properties of 3D braided composites (Bilisik 2013).

Due to the advantageous features of braided composites, they have huge potential for applications in various technical fields, including aerospace engineering, medical field, civil engineering, sports, and so on. Some of the products have already been commercialized and applied in real situations, and many others are still in the research stage. Some examples of applications of braided composites are fan blade containment in commercial aircrafts; energy-absorbing crash structures in racing cars; reinforcement for aircraft propellers and stator vanes in jet engines; lightweight frames and structures such as trusses; reinforcement for drive shafts and torque transfer components, such as flanged hubs; products with changing geometries like prosthetics and hockey sticks; composite rebars for concrete reinforcement; and so on.

4.2 METHODS OF PRODUCTION

4.2.1 THERMOSETTING COMPOSITES

The resin transfer molding (RTM) process is a popular method of manufacturing composite parts with complex shapes due to several advantages such as low cost, low void content, high production rate, and lower risk of exposing the workers to harmful gases and chemicals (Beckwith and Hyland 1998). RTM is frequently used to fabricate braided composite materials based on thermosetting matrix for different applications. The schematic of the RTM process for fabricating a braided composite part is shown in Figure 4.2 (Uozumi et al. 2005). The RTM machine usually contains four units, namely, pumping unit, heating unit, mold unit, and vacuum unit. To start the process, the braided preform is first placed in the mold and closed. The mold is subsequently placed within the heating unit. Due to the action of the pumping and vacuum units, the resin is drawn into the mold and saturates the preform. The temperature of the heating unit is controlled to keep lower resin viscosity and ensure proper curing of the resin. Figure 4.3 shows the I-shaped braided composite part fabricated using the aforementioned process.

It should be noted that for the production of braided composites using the RTM process preforms of desired shapes should be prepared in advance. In case of complex shapes, the preforms are produced by 3D braiding process. To avoid the cost of 3D braiding process, alternatively, preforms can be produced using conventional

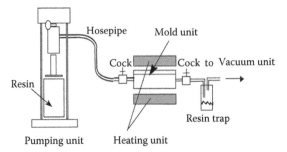

FIGURE 4.2   Schematic diagram of the resin transfer molding process for manufacturing braided composite parts. (From Uozumi et al., *Adv.Compos. Mater.*, 14, 365–83, 2005.)

FIGURE 4.3   I-shaped braided composite fabricated using the resin transfer molding process. (From Uozumi et al., *Adv. Compos. Mater.*, 14, 365–83, 2005.)

2D braiding technique and, subsequently, a deforming process can be carried out to obtain the desired shapes. An example of a deforming process to obtain an I-shaped braided preform is shown in Figure 4.4.

To develop a particular product, braided structures for different parts of the product are first produced, deformed to give the desired shape to each part, and then joined together to give the final shape of the product prior to resin impregnation. The process that can be followed for the production of a box beam, which is used as one of the structural members in wing parts of aircraft, is shown in Figure 4.5. Circular braided preforms can be first produced and deformed to produce the side panel. The bottom panel is produced by combining trapezoid-shaped braids with another braided perform with specific shape produced from a large circular braid

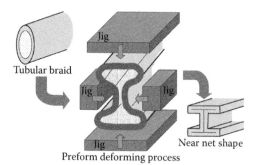

FIGURE 4.4 Deforming process for obtaining I-shaped preforms from tubular braid. (From Uozumi et al., *Adv. Compos. Mater.*, 14, 365–83, 2005.)

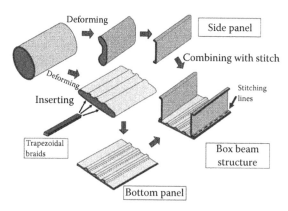

FIGURE 4.5 Production of a box beam preform. (From Uozumi et al., *Adv. Compos. Mater.*, 14, 365–83, 2005.)

by the deforming process. The bottom panel and the side panel are then combined together by the stitching process to get the final shape of the box beam, as shown in Figure 4.6a. The final composite part produced through the RTM process is shown in Figure 4.6b.

However, the process of preform production, deforming, and composite production can be combined in a single process to obtain circular 2D braided composites. A process has recently been developed for the single-step production of braided composite rods (BCRs) using this method (Fangueiro et al. 2011), as schematically shown in Figure 4.7. In this process, axial fibers such as glass, carbon, and so on, are impregnated with a thermosetting polymeric resin and hardener mixture and drawn into a vertical braiding machine. Another set of yarns is drawn from a number of bobbins and braided around the axial fibers by the circular braiding setup. The excess resin is controlled by a ring, and the resin-impregnated preform is then cured in a separate oven to obtain the circular braided composites. In a modified version of this process, a horizontal braiding setup is used and combined with a curing chamber also, to get braided rods in a truly single process.

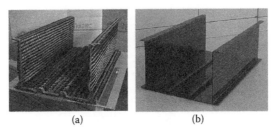

(a)   (b)

FIGURE 4.6 (a) Box beam preform and (b) composite part. (From Uozumi et al., *Adv. Compos. Mater.*, 14, 365–83, 2005.)

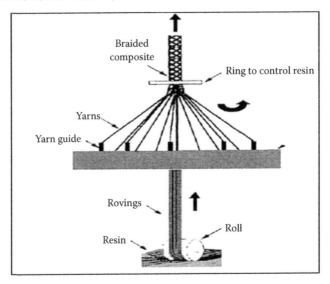

FIGURE 4.7 Schematic diagram of braided composite rod production. (From Fangueiro et al., International Patent 103581, 2011.)

Another similar approach for producing braided composites is to combine the braiding process with a pultrusion line (Ahmadi et al. 2009). This process is schematically shown in Figure 4.8. In this process, axial fibers are drawn from a creel and aligned by passing through an alignment card before entering the resin bath. A series of rollers located in the resin bath spreads and flattens the axial fiber rovings and ensures good wetting of the fibers. The fibers are then passed through a pre-die, which collects and bundles the axial fibers together and gives them a circular cross section. After the braiding of another set of fibers over these axial fibers, the braided structure passes through a secondary resin bath to ensure proper wetting of the fibers with resin. Next, the impregnated braided structure enters the heated pultrusion die, where it gets cured and comes out as a BCR. A pulling mechanism is used to drive the composite rod to the forward direction, where it is cut down to the required length using a saw. Figure 4.9 presents BCRs produced through this method using glass fibers and unsaturated isophthalic polyester resin. For these composites, a pultrusion die 500 mm in length and 6 mm in diameter with temperature of 120°C was used and pultrusion was performed at a speed of 100 mm/min.

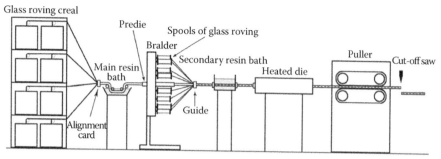

FIGURE 4.8 Schematic diagram of braiding-pultrusion process. (From Ahmadi et al., *Express Polym. Lett.*, 3, 560–8, 2009.)

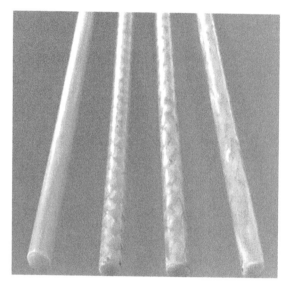

FIGURE 4.9 Braided pultruded rods. (From Ahmadi et al., *Express Polym. Lett.*, 3, 560–8, 2009.)

4.2.2 THERMOPLASTIC COMPOSITES

Techniques similar to braid pultrusion have also been utilized to produce thermoplastic braided composites (Memon and Nakai 2013a,b). For this purpose, it is necessary to mix the braiding yarns with the thermoplastic resin yarns to prepare an intermediate material before the composite manufacturing. Different approaches that can be opted for this purpose are shown in Figure 4.10 (Khindker et al. 2006; Shikamoto et al. 2008; Takai et al. 2006; Tanaka et al. 2009). During the composite fabrication process, the thermoplastic part of the structure melts and forms the matrix of the composites.

A pultrusion molding process of developing jute/polylactic acid (PLA) braided composites has been used by Menon and others (Memon and Asami 2013). In this

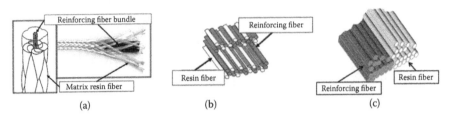

(a)   (b)   (c)

FIGURE 4.10 Schematic of intermediate material: (a) microbraided yarn, (b) comingled yarn, and (c) paralleled yarn. (From Memon and Nakai 2013a.)

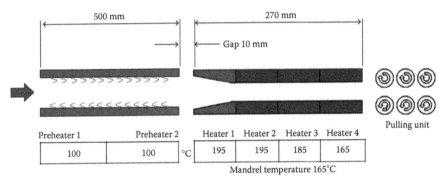

FIGURE 4.11 Schematic diagram of pultrusion molding die used for jute/polylactic acid composites. (From Memon and Nakai 2013b.)

process, jute fiber was first comingled with PLA fibers and the comingled fibers were then braided using a tubular braiding machine. Additionally, glass fibers were used to axially reinforce the structures, to improve the mechanical properties. Pultrusion molding of these structures was then carried out using a tubular die, as shown schematically in Figure 4.11. The braided structure was first passed through a preheater section before entering the pultrusion die. The temperatures of these different heating zones were selected taking care of the melting temperature of PLA (~175°C) and the degradation temperatures of jute and PLA (~240°C and ~320°C, respectively). The pulling mechanism located at the end of the pultrusion mold was used to pull the braided composite at a speed of 18 mm/min. The produced tubular composites are shown in Figure 4.12. It was observed that the selected pultrusion temperature had a strong influence on the quality of the produced tubes. An increase in the pultrusion temperature from 195°C to 225°C led to a considerable decrease in the void convent and unimpregnated area of the composite tubes. However, pultrusion molding at 235°C was not successful for these fibrous structures.

A similar pultrusion molding technique has also been developed to produce thermoplastic (PA 66) braided tubes reinforced with aramid and carbon fibers. In this case, the preheater was maintained at a temperature of 230°C, which was near the melting temperature of polyamide fibers, to achieve easy impregnation, and the four heating zones of the pultrusion die were set at 300°C, 300°C, 275°C, and 230°C.

FIGURE 4.12 Hybrid glass/jute reinforced thermoplastic braided tubes. (From Memon and Nakai 2013b.)

4.3 PROPERTIES OF BRAIDED COMPOSITES

4.3.1 MECHANICAL PROPERTIES

The mechanical properties of braided composites are highly dependent on the structural parameters. Tensile and flexural properties of braided composites improve with the amount of fibers in the axial direction. As a result, triaxial braided composites present higher mechanical performance compared to biaxial braided composites with a similar fiber volume fraction (Potluri et al. 2006). The flexural modulus of braided composites was found to decrease with increase in the braiding angle, due to a decrease in the axial load-sharing fiber component. Similarly, flexural strength was also found to be better with the lower braiding angle due to the same reason. It was observed that braided composites with 31° angle showed fiber breakage as the main failure mechanism, whereas for braided composites with 65° braiding angle the main failure mechanism was matrix cracking due to higher contents of off-axis fibers and high crimp values.

The influence of crimp and alignment of axial fibers on the tensile properties of triaxial braided fabric and composites can be understood from Figure 4.13 (Pereira et al. 2008). In the initial stage, that is, stage II, axial fibers can only support the tensile load partially due to the presence of crimp in the fibers. As crimp is removed, axial fibers share the load in full potential and, therefore, load increases elastically until the failure of axial fibers. Load then drops drastically and braided fibers start to share some load in stage III. It is obvious that the mechanical properties of triaxial braided composites are highly dependent on the type of axial fibers. Using high-strength and high-modulus axial fibers results in improved modulus and strength of braided composites. Figure 4.14 shows the influence of axial fibers on the bending modulus and strength of triaxial braided composites.

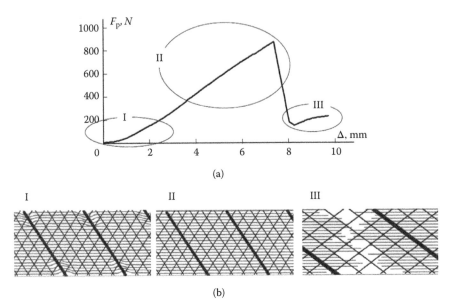

FIGURE 4.13 Deformation mechanism of triaxial braided fabric explained from load-elongation curve (a) and schematic of fibre deformation and rupture within a braided structure (b) at different deformation stages, I, II and III. (From Pereira et al., *Mech. Compos. Mater.*, 44, 221–30, 2008.)

The strong influence of fiber type and hybridization on mechanical properties has also been observed in the case of 3D braided fabric reinforced thermoplastic matrix (nylon) composites (Wan et al. 2005). Braided composites were developed using carbon (30 vol%) and Kevlar fibers (30 vol%) and also by mixing carbon and Kevlar fibers at different proportions. The flexural behavior of these composites is shown in Figure 4.15. It can be observed that all braided composites presented a linear elastic region after which a plateau existed. The composites containing only carbon fibers (C_{3D}/MC) and hybrid composites with a lower amount of Kevlar fibers (20%, sample A) showed linearity after this plateau until the failure of the composites due to mainly breakage of fibers. On the other hand, in case of composites with only Kevlar fibers (K_{3D}/MC) and hybrid composites with a higher amount of Kevlar fibers (sample D), due to existence of other damage mechanisms after this plateau such as interfacial failure, matrix plastic deformation, crack propagation, and so on, load increased with displacement in a nonlinear fashion. No evidence of fiber breakage was observed in either the tensile or the compressive side of the specimens, and fiber buckling was understood as the main failure mode for these composites. The hybrid composites containing the higher amount of Kevlar fibers presented higher breaking strain and ductility and showed the positive hybrid effect.

The positive effect of hybridization on flexural modulus and strength can be seen in Figure 4.16a. It can be observed that hybrid braided composites containing 20% Kevlar fiber showed significantly higher modulus and strength compared to 100% carbon fiber composites. However, presence of very high amounts of Kevlar fiber (80%) deteriorated both strength and modulus of composites. On the contrary, both shear

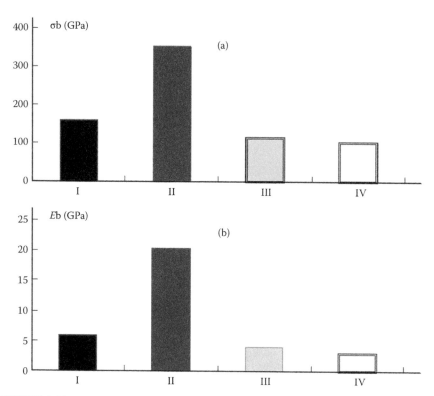

FIGURE 4.14 Influence of axial fiber on (a) bending modulus and (b) strength of braided composites: I, glass; II, carbon; III, high tenacity polyethylene; and IV, sisal. (From Pereira et al., *Mech. Compos. Mater.*, 44, 221–30, 2008.)

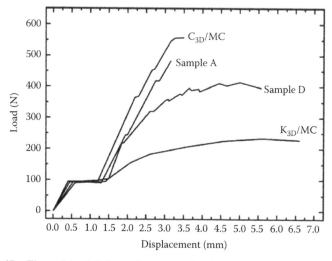

FIGURE 4.15 Flexural load deformation curves for three-dimensional (3D) braided composites. (From Wan et al., *Mater. Sci. Eng. A*, 398, 227–32, 2005.)

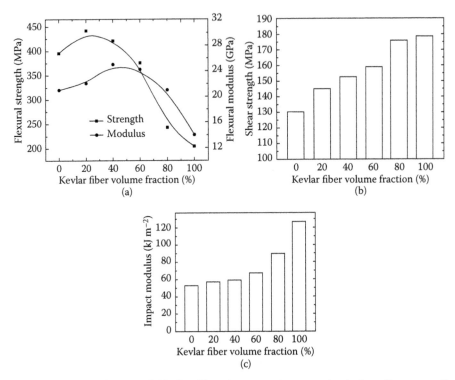

FIGURE 4.16 Influence of Kevlar fiber (percentage) on mechanical performance of 3D hybrid braided composites: (a) flexural strength and modulus, (b) shear strength, and (c) impact modulus. (From Wan et al., *Mater. Sci. Eng. A*, 398, 227–32, 2005.)

strength and impact strength of 3D braided composites improved continuously with the increase in Kevlar fiber percentage. Due to the high energy absorption capability of Kevlar fiber, hybrid composites with higher amounts of Kevlar fiber resulted in higher impact performance. Besides other mechanisms such as plastic deformation of matrix, fiber pull-out, and fiber breakage, the plastic deformation of Kevlar fiber was the main reason enhancing the impact energy absorption capability of 3D braided composites.

4.3.2 SHEAR PERFORMANCE

Braided composites show better shear performance compared to unidirectional composite materials (Ahmadi et al. 2009). Under torsional forces, BCRs show a linear increase in torque with twist angle without any fiber breakage, as shown in Figure 4.17. With subsequent increase in the twist angle, braided cover starts to break and torque increases quasilinearly until the complete rupture of the braided cover occurs. Debonding of interface between braided core and cover then occurs, resulting in a drastic fall in the torque values. On the other hand, unidirectional composite rods having similar fiber volume fractions show inferior shear performance. In this case, after the elastic region torque continues to increase with the twist angle only marginally due to matrix cracking and fiber breakage mechanisms.

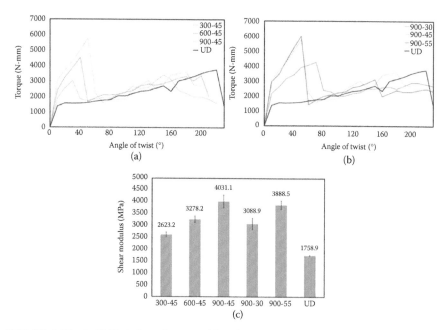

FIGURE 4.17 (a, b) Variation of torque with twist angle and (c) comparison of shear modulus of braided and unidirectional composite rods. (From Ahmadi et al., *Express Polym. Lett.*, 3, 560–8, 2009.)

Because the shear performance is mainly dependent on the braided cover, an increase in the linear density of braided cover (i.e., braided cover to axial fiber ratio) leads to an increase in the shear modulus of braided rods. On the other hand, a braid angle of 45° was found to be optimum for improved shear performance of braided rods. The shear modulus of braided rods with only 7.7 wt% of cover (with respect to axial fibers) was found to be 1.5 times of the unidirectional composite rods having similar fiber volume fractions.

4.3.3 Energy Absorption Capacity

Braided composites present higher energy absorption capability compared to unidirectional fiber reinforced composites (Hamada et al. 2000). Under compressive load, the cracks in the axial fibers can be prevented by the braided fibers, which can apply radial compressive stresses. Figure 4.18 shows the fracture behavior of a unidirectional composite rod and braided rod under compressive forces. It can be observed that the braided rod provided better strength and ductile behavior, resulting in higher energy absorption capability. It was also observed that in the unidirectional rod catastrophic crack propagation occurred along the fibers, whereas progressive crushing was observed in the case of braided rods. The broken parts of the braided rod splayed out radially, as shown in Figure 4.19, and no fracture in the axial fibers was noticed.

Energy absorption capability of braided rods was further improved through a flexible fiber–matrix interface using a flexible type of epoxy resin (Hamada et al. 2000). The braided rod with a flexible interface showed a mean load of over 20 kN, whereas

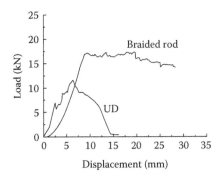

FIGURE 4.18 Compressive behavior of unidirectional (UD) and braided rods. (From Hamada et al., *Compos. Sci. Technol.*, 60, 723–9, 2000.)

FIGURE 4.19 Fracture behavior of (a) braided rod and (b) unidirectional rod. (From Hamada et al., *Compos. Sci. Technol.*, 60, 723–9, 2000.)

the usual braided rod showed 15 kN or less. Energy absorption of the modified rod was 20% higher compared to the usual rods (improved from 69.9 to 83.8 kJ/kg). The higher number of fiber fractures in the case of modified rods was identified as the main reason for the higher energy absorption capability.

The 2.5D thermoplastic braided composite tubes exhibit very good crushing performance comparable to that of metals (Priem et al. 2014). Glass/polypropylene and carbon/polyamide tubes possess specific energy absorption between 36 and 61 kJ/kg, which is comparable or better than steel tubes (between 12 and 38 kJ/kg) and aluminum tubes (22–43 kJ/kg). The crushing performances of different braided tubes having different through-the-thickness bias fiber orientations are provided in Tables 4.1 and 4.2. Carbon/polyamide braided tubes show fragmentation crushing mode characterized by progressive fracturing of composite materials into small pieces and high energy absorption. On the other hand, glass/polypropylene braided tubes show progressive folding behavior similar to metallic tubes (in case of high braiding angles), involving several buckling processes and keeping structural integrity. Splaying is another crushing mode observed in the case of glass/polypropylene tubes at lower braiding angles showing progressive tearing and bending of the tube's wall. Specific energy absorption

TABLE 4.1
Crushing Performance of Glass/Polypropylene Braided Tubes

Samples	Orientation	Impact Angle (°)	F_{max}	$F_{average}$	Specific Energy Absorption (kJ/kg)
1	20	0	81,088	32,153	17,543
2	20	0	85,042	38,324	20,776
3	20	0	79,160	40,816	22,334
4	20	0	108,244	26,681	14,512
5	20	0	83,975	49,387	24,311
6	45	0	96,527	49,452	23,097
7	45	0	106,762	46,873	22,408
8	45	0	87,376	71,351	34,331
9	45	0	143,432	73,377	34,546
10	75	0	73,753	48,379	26,130
11	75	0	67,213	49,482	26,938
12	75	0	67,024	57,511	34,171
13	75	0	71,456	64,329	37,676
14	20	15	40,783	28,237	14,833
15	20	15	46,752	37,040	18,876
16	20	15	29,777	18,493	8,088
17	45	15	69,424	58,031	27,048
18	45	15	68,742	62,910	29,087
19	75	15	55,620	45,883	26,656
20	75	15	58,551	43,484	24,823

Source: Priem et al., *Compos. Struct.*, 116, 814–26, 2014.

TABLE 4.2
Crushing Performance of Carbon/Polyamide Braided Tubes

Samples	Orientation	Impact Angle (°)	F_{max}	$F_{average}$	Specific Energy Absorption (kJ/kg)
1	30	0	140,254	108,671	55,126
2	30	0	151,463	134,794	70,774
3	30	0	146,011	98,174	56,470
4	45	0	145,748	118,716	55,282
5	45	0	100,737	86,081	40,117
6	45	0	131,575	107,758	50,260
7	45	0	142,594	119,446	58,431
8	45	0	129,795	100,311	48,781

Source: Priem et al., *Compos. Struct.*, 116, 814–26, 2014.

of the braided tubes is highly dependent on the braiding angle. Energy absorption increases with braiding angle in the case of glass/polypropylene and decreases in the case of carbon/polyamide braided tubes due to difference in the crushing mode, which is also strongly influenced by the braiding angle. A higher braiding angle leads to lower energy absorption in the case of fragmentation mechanism as this mode involves energy absorption and fracture due to axial compression loads and, therefore, a higher braiding angle reduces axial stiffness and energy absorption capacity. On the other hand, the higher braiding angle in case of glass/polypropylene braided tubes increases the transverse stiffness, making tearing of the tube wall really difficult and facilitating progressive folding and high energy absorption. Therefore, energy absorption is strongly dependent on the type of fiber, matrix, and crushing modes. Additionally, specific energy absorption decreases with the increase in length to diameter ratio and in case of off-axis crushing loads as well.

4.3.4 SURFACE PROPERTIES

Braided composites present a characteristic surface texture due to intertwining of the yarns (Cunha et al. 2014). Figure 4.9 shows the surface of braided rods. The ribbed surface texture can be clearly seen. This type of surface texture can be very useful in improving the interfacial strength of the braided rods with surrounding matrix due to mechanical interlocking. However, the type of surface texture is highly dependent on various parameters such the braiding angle, type of braiding yarns, braiding speed, and so on. Recent studies have shown that the surface texture of braided rods can be easily tailored varying these parameters and it is possible to produce braided rods with excellent bonding with the surrounding matrix. Figure 4.20 schematically shows the cross sections of three types of braided rods produced by varying different parameters, which are listed in Table 4.3. These rods were produced using carbon fibers as axial reinforcements and by braiding polyester yarns around these axial fibers. However, surface tailoring was performed by replacing one or two of these simple polyester yarns with braids made of eight polyester yarns. These braided yarns were much coarser than the simple polyester yarns, and their incorporation into braided rods completely changed the surface texture, as shown in Figure 4.21.

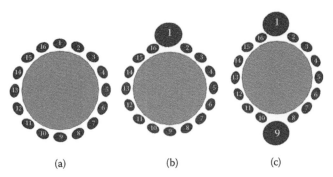

(a)   (b)   (c)

FIGURE 4.20 Cross sections of different types of braided rods: (a) SIMPLE, (b) 1B_LS, and (c) 2B_HS or 2B_LS. (From Cunha et al., *Bull. Mater. Sci.*, 37, 1013–6, 2014.)

TABLE 4.3
Details of Produced Composite Rods

Sample Code	Core Structure	Braided Structure	Speed (m/min)	Braiding Angle (°)
SIMPLE	2 Carbon yarns	16 Polyester yarns	0.54	35
1B_LS	2 Carbon yarns	15 Polyester yarns and 1 braided structure made of 8 polyester yarns	0.54	33
2B_HS	2 Carbon yarns	14 Polyester yarns and 2 braided structures made of 8 polyester yarns	1.07	15
2B_LS	2 Carbon yarns	14 Polyester yarns and 2 braided structures made of 8 polyester yarns	0.54	49

Source: Cunha et al., *Bull. Mater. Sci.*, 37, 1013–6, 2014.

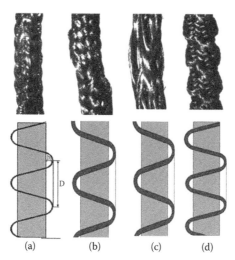

FIGURE 4.21 Surface texture of braided rods: (a) SIMPLE, (b) 1B_LS, (c) 2B_HS, and (d) 2B_LS. (From Cunha et al., *Bull. Mater. Sci.*, 37, 1013–6, 2014.)

It can be observed that the change in the braiding parameters resulted in a change in h (height of surface rib) or D (distance between adjacent ribs) or both. When these braided composites were incorporated within a cementitious matrix, the surface texture exhibited strong influence on the bonding characteristics. Among these various rods, the one produced using two braids and 14 simple polyester yarns and at a higher take-up speed of 1.07 m/min exhibited excellent bonding characteristics. This type of braided rods presented complete tensile rupture before debonding and, therefore, their tensile strength was fully utilized.

4.3.5 Frictional and Wear Behaviors

Frictional and wear behaviors of 3D braided carbon/epoxy composites under dry sliding conditions have been investigated (Wan et al. 2006). It was observed that the friction and wear characteristics were dependent on different material parameters such as fiber volume fraction and fiber–matrix interface and testing parameters such as normal load and sliding velocity. The wear mechanism observed in the case of carbon/epoxy 3D braided composites was similar to that of conventional fiber rein-forced composites. The specific wear rate decreased steadily with the increase in fiber volume fraction, as shown in Figure 4.22. Abrasion was noticed as the primary wear mechanism in the case of carbon/epoxy 3D braided composites. The worn sur-face of the composites showed exposed fiber ends and wear debris, as shown in Figure 4.23. The exposed fiber ends helped to support a great portion of the applied load, reducing the direct interaction between the matrix and the metallic abrasive part. This resulted in protection of matrix materials by reinforcing fibers, leading to a lower wear rate of the composites. Besides fiber volume fraction, wear behavior of these composites was also influenced strongly by the fiber–matrix interface. Similar to conventional composites, improved fiber–matrix interface reduced the fiber–matrix debonding, keeping the broken fibers on the composite surface and thereby preventing the formation of third-body abrasives in earlier stages.

Besides material parameters, test parameters such as sliding distance also has a strong influence on the coefficient of friction and wear behavior of 3D braided composites (Wan et al. 2007). Under both dry and lubricated sliding conditions, fractional and wear behaviors of 3D carbon/Kevlar/epoxy braided composites are similar to those of conventional composite materials. With increase in the sliding distance, coefficient of friction and wear rate both decrease considerably and then level off. This behavior is not influenced by the hybridization. The coefficient of friction and wear rate are, however, dependent on the normal load and sliding con-ditions. The dry sliding condition results in a much lower friction coefficient and

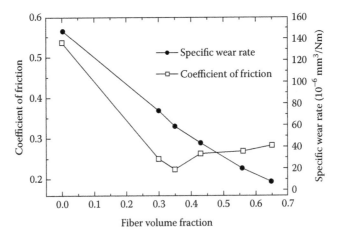

FIGURE 4.22 Influence of fiber volume fraction on wear behavior of 3D braided compos-ites. (From Wan et al., *Wear*, 260, 933–41, 2006.)

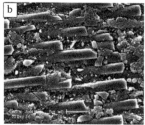

FIGURE 4.23 Worn surfaces of 3D carbon/epoxy braided composites: (a) $V_f = 0.43$ and (b) $V_f = 0.65$. (From Wan et al., *Wear*, 260, 933–41, 2006.)

specific wear rate. For the 3D braided composites, hybridization shows negative effects on friction and wear behaviors and, among the different hybrid composites, the one with a carbon to Kevlar ratio of 3.2 shows the best wear resistance and lowest frictional coefficient.

4.4 LATEST DEVELOPMENTS

4.4.1 Braided Composites for Health Monitoring

Recently, braided composites were developed for various applications where safety of human beings is a prime factor. Examples of such applications are aerospace or civil engineering structures. Catastrophic failure of such structures can lead to huge damage to human lives. Braided composites used as reinforcement for these structures can be used as a health monitoring system, to detect the damages well before the structural collapse. One such attempt has been made to develop a strain-sensing braided composite incorporating optical fiber sensors (Kosaka et al. 2004). A fiber-optic sensor was embedded within the braided composite during resin impregnation, as shown in Figure 4.24.

It was observed that the fiber-optic sensor embedded within the braided composite could monitor the residual thermal strain during the curing process. Moreover, it was observed that the used sensor could monitor effectively the strain level under tensile loading, as shown in Figure 4.25. It is clear that the sensor can monitor tensile strain accurately up to 0.8% strain, after which slight error in the detected strain level was observed due to initiation of damage in the composite structure. In case of fatigue loading, the measured strain from fiber-optic sensor followed the same cyclic behavior as that measured from strain gauges below 15,000 cycles. However, for 15,000 cycles or higher the two strain peaks obtained from the fiber-optic sensor suggested the initiation of damage in the composite structure such as matrix cracking. Therefore, it was possible to detect the initiation of damage in the braided composites using fiber-optic sensors.

Although fiber-optic sensors have been widely investigated for health monitoring and are available commercially, high cost, fragile nature, complex operation, and so on, are the disadvantages associated with these sensors (Rana et al. 2014). Therefore, there exists a strong need for simple and low-cost solutions for health monitoring. Self-sensing composites have come up as one of the alternative materials

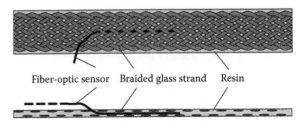

FIGURE 4.24 Incorporation of fiber-optic sensor within a braided composite. (From Kosaka et al., *Adv. Compos. Mater.*, 13, 157–70, 2004.)

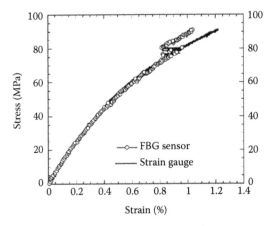

FIGURE 4.25 Strain sensing in braided composites using fiber Bragg grating (FBG) sensor. (From Kosaka et al., *Adv. Compos. Mater.*, 13, 157–70, 2004.)

for strain and damage sensing of various structures. Principally, these composites contain a conductive element in their structure such as carbon fiber. When subjected to mechanical deformation, a change occurs in the electrical contact points, leading to a change in the electrical resistivity of composites. Under small strain, the resistance of composites changes reversibly as the electrical contacts restore to their original position on removal of the mechanical loading. However, at high strain levels damage in the composite structure such as matrix cracking, fiber breakage, and so on leads to permanent changes in the electrical contacts and as a consequence resistance changes irreversibly to a higher extent. Unidirectional carbon fiber composites show very small change in resistance with mechanical strain, and therefore they are not suitable for health monitoring. Hybrid composites containing a mixture of carbon and glass fiber can be designed in such a way that the composites show a significant change in electrical resistance well before the complete breakage of the composite and, therefore, can be used to generate alarm signals well before complete structural damage (Nanni et al. 2006, 2007).

Recently, it was observed that BCRs containing carbon fibers for axial reinforcement can provide good sensitivity to low strain levels (Rana et al 2014; Rosado et al. 2013). In case of these composites, it was possible to introduce some degree

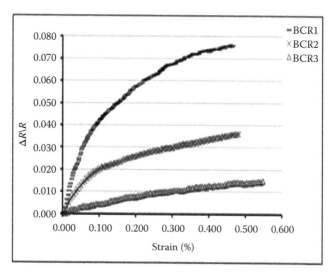

FIGURE 4.26 Change of fractional resistance with mechanical deformation for braided composite rods (BCRs). (From Rana et al., *Sci. World J.*, 2014, 1–9, 2014.)

of misalignment in axial carbon fibers due to overbraiding of the other set of yarns. The misalignment of the axial fibers could be controlled by controlling the speed of the braiding machine and also the tension in the axial fibers. The misaligned arrangement of carbon fibers probably led to more change in electrical contact points during mechanical deformation and, as a result, significant change in electrical resistance was noticed, as shown in Figure 4.26. As the change in electrical contacts is more obvious when a lower amount of carbon fibers is present (less touching of fibers), BCRs with lower amounts of carbon fibers showed higher change in electrical resistance. Under cyclic loading, these composites showed reversible change in resistance with mechanical strain, making them suitable for continuous strain monitoring.

However, conventional composites cannot sense microscale damages in their structure. To overcome this, multiscale composites combining macro- and nanoscale reinforcements have been developed. Nanoscale reinforcement such as carbon nanofiber (CNF) or carbon nanotube (CNT) has the ability to form conductive paths at the nano scale, which can be broken by the initiation of microscale damages, resulting in detectable change in electrical resistance of composites. The 3D braided composites containing CNTs within the matrix have been developed and found useful for detecting microscale damages (Gao et al. 2009; Kim et al. 2010; Thostenson and Chou 2006). Figure 4.27 shows the change in electrical resistance with strain of these multiscale braided composites. The microscale damages such as transverse cracks or microdelamination starts at stage 2 and accumulates in stage 3, leading to significant rise in electrical resistance change. Stage 4 corresponds to the saturation of microdamages, and stage 5 represents the closing of microdamages due to Poisson's contraction and jamming of yarns. Therefore, it is possible to detect microscale damages using multiscale braided composites.

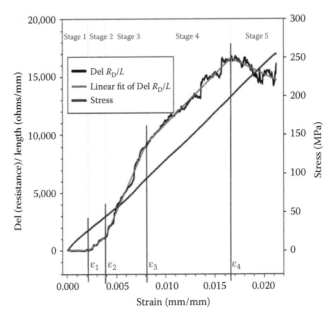

FIGURE 4.27 Change of electrical resistance with strain for 3D braided composites. (From Kim et al., *Compos. Part A*, 41, 1531–7, 2010.)

4.4.2 Multiscale Braided Composites

Although braided composites present several advantageous features, these materials have lower in-plane mechanical properties compared to conventional laminates. The in-plane mechanical properties of braided composites can be improved through the recently developed multiscale reinforcement approach (Bhattacharyya et al. 2013; Rana et al. 2009; Rana et al. 2011a,b, 2012, 2013). This concept involves the enhancement of matrix properties using another set of reinforcements from the nano scale, and the resulting composites are known as multiscale composites. In case of laminated composites, several types of nanoscale reinforcements have already been tried, such as CNTs, CNFs, nanoclays, nano Al_2O_3, and so on, with promising results. Recently, the same concept has also been applied to braided composites using nanoclays. Biaxial and triaxial braided composite laminates have been developed using epoxy resin–containing nanoclays (Hosur et al. 2010). For this purpose, nanoclays were mixed with the epoxy resin using a magnetic stirrer and subsequently used to impregnate the braided preforms using vacuum-assisted resin infusion technique.

It was observed that the braided composites containing 1 wt% of nanoclay exhibited significantly higher flexural modulus and strength compared to control samples without nanoclays. The improvement of flexural strength and modulus were 11% and 33% for biaxial braided composites, 6.6% and 15.3% for triaxial 0°/±45° braided composites, and 20% and 12.4% for triaxial 0°/±60° braided composites, respectively. The improvement was attributed to enhanced mechanical properties of matrix as well as improved interface between fiber and epoxy matrix due to nanoclay addition.

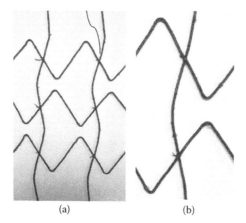

(a) (b)

FIGURE 4.28 Auxetic structure made from (a) braided composite rod and (b) magnified structure. (From Subramani et al., *Mater. Design*, 61, 286–95, 2014.)

In addition to the flexural properties, the addition of nanoclay to the braided composites significantly improved the low velocity impact resistance of the composites. The damage area due to impact at different energy levels was significantly lower in the case of multiscale braided composites compared to the control samples. These promising results from multiscale braided composites suggest a good future for these new-generation composite materials.

4.4.3 AUXETIC STRUCTURES FROM BRAIDED COMPOSITES

One of the primary advantages of composite materials is the design flexibility. Different designs in terms of fiber system, matrix system, or interface may lead to completely different sets of properties. Another approach to obtain a unique set of properties is to combine different composite materials in such a way that the produced structure can provide completely new behavior. One such effort with braided composites has been made recently to produce auxetic structures. In contrast to conventional materials, auxetic materials expand in transverse direction when a tensile force is applied in the longitudinal direction, resulting in a negative Poisson's ratio. Owing to this behavior, these materials possess higher strength and energy absorption, improved fracture toughness, higher damping performance and indentation resistance, and so on (Subramani et al. 2014). Owing to high strength and energy absorption, auxetic structures made from BCRs have been proposed for application in strengthening of masonry walls. These auxetic structures are expected to perform better than the conventional composite laminates and strips used for this purpose. Figure 4.28 shows the type of structure produced from braided rods, and their auxetic behavior is shown in the graph presented in Figure 4.29. It can be clearly seen that these structures provide a very high negative Poisson's ratio (up to –5.20) at low strain and as the strain increases the auxetic behavior decreases considerably. Studies are under way to explore these interesting structures in construction for strengthening masonry walls and concrete structures.

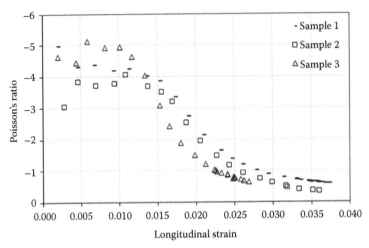

FIGURE 4.29 Poisson's ratio of braided rod–based auxetic structures. (From Subramani et al., *Mater. Design*, 61, 286–95, 2014.)

4.5 SUMMARY AND CONCLUDING REMARKS

Braided composites are produced from braided textile preforms having various shapes and architectures. Due to the possibility of producing near net shapes as well as complex geometries, combined with enhanced transverse and torsional properties and energy absorption capability, braided composites have the potential for application in various industrial sectors including aerospace, medical field, civil engineering, sports, and so on. Some of the applications have been already commercialized, whereas several types of braided composite products are still in the research stage.

RTM is the most commonly used method of producing braided composites due to its cost-effectiveness, higher production, and capability to produce complex composite parts with very less defects. However, other techniques such as hand layup followed by consolidation using a vacuum bag or compression molding can also be used. Usually, the braided preforms are given the desired shape using a deforming process prior to resin application and curing. Alternatively, BCRs can be produced using a single-step braiding process capable of performing preform manufacturing, resin impregnation, and curing in a single step or combining braiding machine with a pultrusion line. In this case, the process speed and temperature of the curing unit or pultrusion die should be properly selected according to the type of materials to produce well-cured and defect-free composites. Thermoplastic braided composites can be developed using braided preforms produced from hybrid yarns containing both reinforcing and matrix filaments and by subsequently melting and consolidating the thermoplastic matrix filaments using a pultrusion die. Hybrid yarns are produced using the comingling technique.

The mechanical properties of braided composites are highly dependent on the type of fibers and the braiding angle. The failure behavior and mechanical properties of braided composites can be greatly enhanced using axial fibers and optimizing the braiding angle. Usually, a lower braiding angle results in higher in-plane mechanical properties due to better load sharing. In case of braided composites with axial fibers, the in-plane mechanical performance is mainly dependent on the type or volume fraction of axial fibers and hybridization of axial fibers may result in better strength as well as stiffness compared to single fiber types due to positive hybrid effect. On the contrary, shear performance of braided composites is mainly dependent on the braided component and also on the braiding angle. Compared to unidirectional composite rods, braided rods show better shear performance. The same has been observed in the case of crushing performance. The crushing energy absorption is significantly higher in braided rods compared to unidirectional composite rods. Progressive crushing behavior is observed in case of braided rods, resulting in excellent energy absorption, and several mechanisms are involved such as splaying, buckling, and fragmentation depending on the type of fiber, matrix, and braiding angle.

For different applications, braided composites are incorporated within another type of matrix as reinforcements and the interfacial strength between the braided composites and the matrix is extremely important. Braided composites usually present a ribbed surface texture, which is helpful to provide a better interface due to mechanical interlocking. Further, the surface can be tailored by just varying the braiding process parameters and it is possible to develop braided composites showing excellent adhesion with other matrices such as concrete, soil, and so on. Fractional and wear behaviors of braided composites are similar to those of conventional composites and are mainly dependent on the fiber and matrix systems, fiber volume fraction, and fiber–matrix interface.

Multifunctional braided composites capable of performing smart functions such as monitoring of strain and damage are developed by either incorporating sensors (such as fiber optic) within the composite structure or turning the composites self-sensing by incorporating conducting fibers or a mixture of conducting and nonconducting fibers within the structure of braided composites.

Application of nanotechnology in composite materials is a recent trend to develop advanced and improved materials. Multiscale braided composites containing CNTs dispersed within the matrix are highly suitable for sensing microscale damages. In addition, dispersion of high-performance nanostructures such as CNT, nanoclay, nanofiber, and so on, may improve the in-plane mechanical performance of braided composites. Some initial studies have shown very promising results, and further research should be conducted in this direction to develop advanced braided composites with multifunctional properties for many existing and future applications.

REFERENCES

Ahmadi, M. S., Johari, M. S., Sadighi, M., and Esfandeh, M. 2009. An experimental study on mechanical properties of GFRP braid-pultruded composite rods. *eXPRESS Polymer Letters* 3: 560–8.

Beckwith, S. W., and Hyland, C. R. 1998. Resin transfer molding: a decade of technology advances. *SAMPE Journal* 34: 7–19.

Bhattacharyya, A., Rana, S., Parveen, S., Fangueiro, R., Alagirusamy, R., and Joshi, M. 2013. Mechanical and thermal transmission properties of carbon nanofibre dispersed carbon/phenolic multi-scale composites. *Journal of Applied Polymer Science* 129: 2383–92.

Bilisik, K. 2013. Three dimensional braiding for composites: a review. *Textile Research Journal* 83: 1414–36.

Cagri, A., and Carey, J. 2008. 2D braided composites: a review for stiffness critical applications. *Composite Structures* 85: 43–58.

Cunha, F., Rana, S., Fangueiro, R., and Vasconcelos, G. 2014. Excellent bonding behaviour of novel surface-tailored fibre composite rods with cementitious matrix. *Bulletin of Materials Science* 37: 1013–6.

Fangueiro, R., Pereira, C., Aaraujo, M., and Jalali, S. 2011. International Patent 103581.

Gao, L., Thostenson, E. T., Zhang, Z., and Chou, T. 2009. Sensing of damage mechanisms in fiber-reinforced composites under cyclic loading using carbon nanotubes. *Advanced Functional Materials* 19: 123–30.

Hamada, H., Kameo, K., Sakaguchi, M., Saito, H., and Iwamoto, M. 2000. Energy-absorption properties of braided composite rods. *Composites Science and Technology* 60: 723–9.

Hosur, M. V., Islam, M. M., and Jeelani, S. 2010. Processing and performance of nanophased braided carbon/epoxy composites. *Materials Science and Engineering: B* 168: 22–9.

Khindker, O. A., Ishiaku, U. S., Nakai, A., and Hamada, H. 2006. A novel processing technique for thermoplastic manufacturing of unidirectional composites reinforced with jute yarns. *Composites Part A* 37: 2274–84.

Kim, K. J., Yu, W., Lee, J. S. et al. 2010. Damage characterization of 3D braided composites using carbon nanotube-based in situ sensing. *Composites Part A* 41: 1531–7.

Kosaka, T., Kurimoto, H., Osaka, K. et al. 2004. Strain monitoring of braided composites by using embedded fiber-optic strain sensors. *Advanced Composite Materials* 13(2004): 157–70.

Memon, A., and Nakai, A. 2013a. Mechanical properties of jute spun yarn/PLA tubular braided composite by pultrusion molding. *Energy Procedia* 34: 818–29.

Memon, A., and Nakai, A. 2013b. The processing design of jute spun yarn/PLA braided composite by pultrusion molding. *Advances in Mechanical Engineering* 2013: 1–8.

Mouritz, A. P., Bannister, M. K., Falzon, P. J., and Leong, K. H. 1999. Review of applications for advanced three-dimensional fibre textile composites. *Composites Part A* 30: 1445–61.

Nanni, F., Auricchio, F., Sarchi, F., Forte, G., and Gusmano, G. 2006. Self-sensing CF-GFRP rods as mechanical reinforcement and sensors of concrete beams. *Smart Materials and Structures* 15.1(2006): 182.

Nanni, F., Auricchio, F., Sarchi, F., Forte, G., and Gusmano, G. 2007. Design, manufacture and testing of self-sensing carbon fibre–glass fibre reinforced polymer rods. *Smart Materials and Structures* 16: 2368.

Pereira, C. G., Fangueiro, R., Jalali, S., Araujo, M., and Marques, P. 2008. Braided reinforced composite rods for the internal reinforcement of concrete. *Mechanics of Composite Materials* 44: 221–30.

Potluri, P., Manan, A., Francke, M., and Day, R. J. 2006. Flexural and torsional behaviour of biaxial and triaxial braided composite structures. *Composite Structures* 75: 377–86.

Priem, C., Ramzi, O., Patrick, R., and Damien, G. 2014. Experimental investigation of the crash energy absorption of 2.5 D-braided thermoplastic composite tubes. *Composite Structures* 116: 814–26.

Rana, S., Alagirusamy, R., and Joshi, M. 2009. A review on carbon epoxy nanocomposites. *Journal of Reinforced Plastics and Composites* 28: 461–87.

Rana, S., Alagirusamy, R., and Joshi, M. 2011a. Development of carbon nanofibre incorporated three phase carbon/epoxy composites with enhanced mechanical, electrical and thermal properties. *Composites Part A* 42: 439–45.

Rana, S., Alagirusamy, R., and Joshi, M. 2011b. Single-walled carbon nanotube incorporated novel three phase carbon/epoxy composite with enhanced properties. *Journal of Nanoscience and Nanotechnology* 11: 7033–6.

Rana, S., Alagirusamy, R., and Joshi, M. 2012. *Carbon Nanomaterial Based Three Phase Multi-Functional Composites*. Saarbrücken, Germany: Lap Lambert Academic Publishing.

Rana, S., Bhattacharyya, A., Parveen, S., Fangueiro, R., Alagirusamy, R., and Joshi, M. 2013. Processing and performance of carbon/epoxy multi-scale composites containing carbon nanofibres and single walled carbon nanotubes. *Journal of Polymer Research* 20:1–11.

Rana, S., Zdraveva, E., Pereira, C. G., Fangueiro, R., and Correia, A. G. 2014. Development of hybrid braided composite rods for reinforcement and health monitoring of structures. *The Scientific World Journal* 2014: 1–9.

Rosado, M. K. P., Rana, S., and Fangueiro, R. 2013. Self-sensing hybrid composite rod with braided reinforcement for structural health monitoring. *Materials Science Forum* 7302: 379.

Shikamoto, N., Wongsriraksa, P. Ohtani, A., Leong, Y. W., and Nakai, A. 2008. Processing and mechanical properties of jute reinforced PLA composite. *Proceeding of Society for the Advancement of Material and Process Engineering conference (SAMPE)*.

Subramani, P., Rana, S., Oliveira, D.V., Fangueiro, R., and Xavier, J. 2014. Development of novel auxetic structures based on braided composites. *Materials & Design* 61: 286–95.

Takai, Y., Kawai, N., Nakai, A., and Hamada, H. 2006. Mechanical properties of long-fiber reinforced thermoplastic composites. *Proceedings of the 12th U.S. Japan Conference on Composite Materials*, The University of Michigan-Dearborn, Michigan, 21–22 September, 2006.

Tanaka, Y., Shikamoto, N., Ohtani, A., Nakai, A., and Hamada, H. 2009. Development of pultrusion system for carbon fiber reinforced thermoplastic composite. Paper presented at 17th International Conference on Composite Materials, Edinburgh, United Kingdom, 27–31 July 2009.

Thostenson, E. T., and Chou, T. W. 2006. Carbon nanotube networks: sensing of distributed strain and damage for life prediction and self-healing. *Advanced Materials* 18: 2837–41.

Uozumi, T., Kito, A., and Yamamoto, T. 2005. CFRP using braided preforms/RTM process for aircraft applications. *Advanced Composite Materials* 14: 365–83.

Wan, Y. Z., Chen, G. C., Huang, Y. et al. 2005. Characterization of three-dimensional braided carbon/Kevlar hybrid composites for orthopedic usage. *Materials Science and Engineering: A* 398: 227–32.

Wan, Y. Z., Chen, G. C., Raman, S. et al. 2006. Friction and wear behavior of three-dimensional braided carbon fiber/epoxy composites under dry sliding conditions. *Wear* 260: 933–41.

Wan, Y. Z., Huang, Y., He, F., Li, Q. Y., and Lian, J. J. 2007. Tribological properties of three-dimensional braided carbon/Kevlar/epoxy hybrid composites under dry and lubricated conditions. *Materials Science and Engineering: A* 452: 202–9.

Zsigmond, B. 2005. "Modeling of braided fibre reinforced composites crosslinked by electron beam." PhD Diss., Budapest University of Technology and Economics, Budapest, Hungary.

Guodong Fang and Jun Liang

CONTENTS

ABSTRACT

An effective way to study the mechanical properties of braided composites is using finite element method (FEM), in which the complex mesostructures of braided composites can be taken into account. In this chapter, the main mesogeometry modeling and mesh generation methods for braided composites are discussed. Discussions presented here shows that based on the periodic characteristic of braided composites, the representative unit cell (RUC) can be chosen to study effective properties, progressive damage development, and failure analysis. The material properties of braiding yarn can also be calculated by the basic material properties of reinforced fiber and matrix. The damage modes, such as braiding yarn breakage, yarn–matrix interface debonding, and matrix cracking, within braided composites can be modeled by numerical models. The biaxial loading response and the failure loci of braided composites can also be studied. Finally, the limitations of the numerical method and the development perspective for analysis of braided composites are summarized.

5.1 INTRODUCTION

It is essential to analyze the mechanical properties of textile composites before using them for a wide range of engineering applications. However, owing to braided composites having complicated textile structures, anisotropic properties of constituents, and their nonhomogeneous nature, it is a challenging task to precisely evaluate the mechanical properties of braided composites, especially for strength analysis. There exist some limitations to apply braided composites to reach the level of implementation of laminated composites (Pastore 2000). In recent years, considerable attention was paid to model braided composites or other textile composites based on some assumptions. Usually, a RUC of textile composites is adopted to predict their mechanical properties due to the periodic characteristics of the microstructure of textile composites, not excepting braided composites. The RUC chosen to epitomize the entire material is the smallest subvolume (Aboudi 1991). It should be noted that the periodic characteristics may be satisfactory for components with regular shape, but not for components with curved shape or holes. However, the RUC analysis method can not only consider the geometric characteristics for each textile composite in detail, but also greatly decrease the computational complexity. At present, the RUC method is still an effective tool to study the stiffness properties, stress–strain fields, damage development, and failure analysis of braided composites. Two aspects should be taken into account when mechanical properties of braided composites are evaluated (Fang and Liang 2011). First, the geometry of braided composites can be divided into three levels: microlevel, mesolevel, and macrolevel. They correspond to the scale of fiber or filament, braiding yarn, and the whole structure, respectively. Therefore, the analysis of braided composites is a multiscale problem. It is an important issue to construct the relationship among the different scales. When the material parameters of constituents (such as filament) in the microlevel are known, the mechanical properties of braided composites in the macrolevel could be predicted. On the contrary, when the boundary conditions and loading of braided composites in the macrolevel are given, the response (stress and strain) of the constituents could be obtained. Then, the elastic properties, damage, and failure of braided composites in the different levels could be predicted and evaluated. However, the geometry of braided composites should be described in detail, which may directly influence the accuracy of the prediction model. Usually, the key influencing factors, such as fiber orientation and fiber volume fraction, for the mechanical properties of braided composites should be equivalent to those of real braided composites. Furthermore, the interface bonding state and initial damage or voids could be included in the prediction model.

Different theoretical, numerical, and experimental methodologies have been developed to investigate the mechanical properties of braided composites. This chapter mainly focuses on the geometry modeling and numerical analysis of braided composites. These numerical analyses can also be used for other textile composites with periodic structural characteristics.

5.2 MESOGEOMETRY MODELING AND MESH GENERATION

5.2.1 Mesogeometry Modeling

A realistic and elaborate mesoscopic geometry modeling of braided composites, considering braid architecture, orientation and distribution of yarns, yarn sizes, braid pitch length, and fiber volume fraction, and so on, is essential and the preliminary task to predict valid and precise material properties by numerical methods. There are several methods to obtain the configuration of textile composites. First, the internal yarn path and cross sections of braided composites are obtained by cutting braided composite samples (Wang and Wang 1995a; Chen, Tao, and Choy 1999a). Second, optical microscopy and micro–computed tomography (μCT) scanning, which have the benefit of being nondestructive, can also be conducted to obtain cross-sectional images of a composite material. μCT is based on differences in x-ray attenuation of materials, which are related to the density of material discriminating the interior microstructure (Stock 1999). The full information about the interior geometry of textiles obtained by μCT can be used as input in conjunction with a textile software tool. There are several virtual textile composite software tools, such as WiseTex (Lomov and Verpoest 2003; Verpoest and Lomov 2005; Straumit et al. 2013) and TexGen (Sherburn 2007), which are the integrated textile preprocessing tools for mesomechanical, hydrodynamical, and structural analysis software packages. Last, textile preforms could be constructed to simulate the textile manufacturing process by using a digital element model (Wang and Sun 2001; Sun 2004). A braiding yarn was modeled by a pin-connected digital-rod-element chain. The length of the element approached approximately zero. Therefore, the yarn was flexible and similar to the true braiding yarn used in the braiding process, but it could not have variation in cross section. To overcome this weak point, a multichain digital element model is used to simulate the irregular cross section of yarn (Zhou, Sun, and Wang 2004; Miao et al. 2008; Green et al. 2014).

In this chapter, we mainly focus on the geometry of three-dimensional (3D) four-step braided composites. It can be found from the mesoscopic yarn configuration of braided composites that the microscopic structures in the surface and corner regions are different from the structure in the interior regions. Three-cell models, as shown in Figure 5.1, using corner, surface, and interior RUCs are studied by some scholars (Wu 1996; Byun and Chou 1996). Usually, the path of yarns within rectangular braided composites is considered to be straight and aligned. And the cross section of braiding yarn can be assumed to be circular, elliptical, hexagonal, and octagonal (Chen, Tao, and Choy 1999a; Zeng, Wu, and Guo 2004; Liu, Lu, and Yang 2006; Fang, Liang, and Wang 2009). If the cross section of braiding yarn is assumed to have an elliptical shape, the long and short axes are a and b, respectively. Meanwhile, the braiding process is considered to be stable to obtain uniform braid structures. If the three cells are piled together, the geometry relations among the internal cell, surface cell, and corner cell can be obtained. After observation and computation, the interior braid angle, γ; surface braid angle, θ; corner braid angle, β; and braid angle of the braided composites, α, have the following relationship:

$$\tan\gamma = \sqrt{2}\tan\alpha = \frac{12}{\pi\tan\theta} = \frac{12}{\pi\tan\beta} \tag{5.1}$$

Figure 5.2 shows the microstructures of 3D four-step braided preforms in a section obtained by cutting longitudinally at 45° angle with respect to the preform

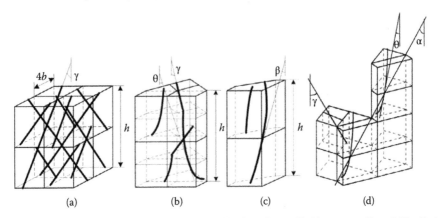

FIGURE 5.1   Three-cell model: (a) interior cell, (b) surface cell, (c) corner cell, and (d) piled together.

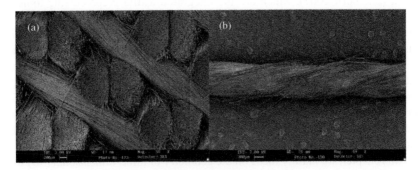

FIGURE 5.2   Simple electron microscope micrographs of three-dimensional four-directional braided composites: (a) the interior of a four-step braided preform and (b) a braiding yarn in the preform interior.

surface. The smallest RUC for the interior structure of a braided composite is shown in Figure 5.3. This can reproduce the geometrical structure of braided composites when RUC is piled up periodically and could be extracted from the periodic structures. Figure 5.3c shows the dimensions of cross-section of braiding yarn located in the RUC. The braid angle γ and the height h of the RUC can be measured by microscopic image analysis. The relations of geometrical parameters indicated in Figure 5.3b and c of the RUC can be expressed as follows:

$$h = 8b/\tan\gamma \tag{5.2}$$

$$L_a = 2b\cos\gamma \tag{5.3}$$

$$L_b = 2b - (2a - L_a)\cos\gamma \quad L_b \leq 2a \leq L_m \tag{5.4}$$

$$L_m = 4b\cos\gamma \tag{5.5}$$

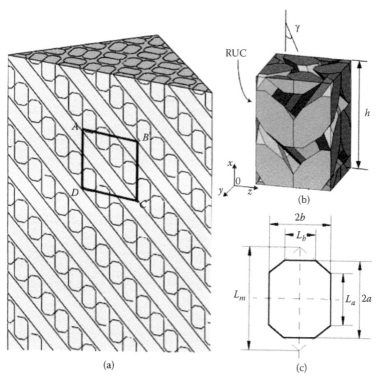

FIGURE 5.3 Representative unit cell (RUC) and its geometrical dimensions for three-dimensional four-step braided composites: (a) cutting cross section of the braided composites and finding RUC, (b) RUC, and (c) cross section of the braiding yarn.

where the height of cross section $2a$ can be determined by K-number (1000 fibers are counted as 1 K) of yarn and cross-sectional area $A = 6b^2 \cos\gamma - (L_m - 2a)^2/(2\cos\gamma)$. All geometrical parameters of 3D four-step braided composites are given in Table 5.1.

It should be noted that the realistic geometrical shape of yarn would be irregular and undulated due to the squeezing of yarns by each other. As shown in Figure 5.2, the mesostructure of interior braiding yarn was observed by peeling a braiding yarn from the interior structure of braided preforms. It was found that the braiding yarns remain straight at the interior but distort at the surfaces. And the cross sections of twisted yarns become irregular. As the braiding yarns remain under tension in the braiding process, they will contact or even become jammed with each other. For the RUC of 3D four-step braided composites, the arrangement of unidirectional yarns of a RUC is shown in Figure 5.4a. After several yarns are pieced together, it appears as an intact yarn in the RUC, as shown in Figure 5.3b. It can be found that the appearance of squeezing regions among yarns in different directions is periodic. The length of one period is denoted by L, with the following geometrical relation:

$$L = 4b/\sin\gamma \qquad\qquad (5.6)$$

To evaluate the mechanical properties of yarn properly, each yarn in the RUC is divided into seven regions (Fang et al. 2009). The regions and local coordinates in

---

TABLE 5.1

Geometrical Parameters of RUC

Interior Braided Angle, γ (°)	Height of RUC, h (mm)	Fiber Volume Fraction of RUC, V_f (%)	b (mm)	K, Number of Braiding Yarn
45	3.04	51.76	0.38	12
30	5.265	50.9	0.38	12

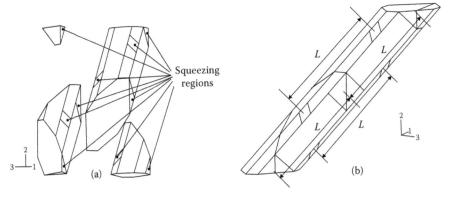

FIGURE 5.4 Braiding yarns in the interior of representative unit cell (RUC): (a) one-directional (1D) braiding yarns in the RUC and (b) an intact braiding yarn pieced together by 1D braiding yarns in the RUC.

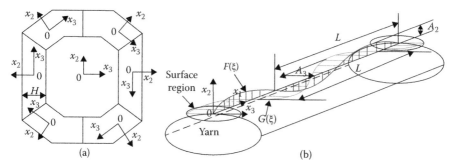

FIGURE 5.5 Cross section and description function of yarn surface fiber path: (a) cross section of braiding yarn and (b) two periodic functions used to describe the fiber path on the braiding yarn surface.

the cross section of the divided yarn are shown in Figure 5.5a. The path of yarn in each region, $\mathbf{r}(\xi)$, can be described by the functions $F(\xi)$ in x_1–x_2 plane and $G(\xi)$ in x_1–x_3 plane, as shown in Figure 5.5b. It can be expressed as follows:

$$\mathbf{r}(\xi) = F(\xi)\mathbf{x}_2 + G(\xi)\mathbf{x}_3 \tag{5.7}$$

where ξ is a variable in x_1 direction.

Because the squeezing regions are periodic, $F(\xi)$ and $G(\xi)$ are periodical functions as well. They can be expressed as

$$F(\xi) = A_2 \cos(\frac{2\pi}{L}\xi - \psi_2) \tag{5.8}$$

$$G(\xi) = A_3 \cos(\frac{2\pi}{L}\xi - \psi_3) \tag{5.9}$$

where, ψ_2 and ψ_3 are phase angles. The variables A_2 and A_3 are the amplitude values of $F(\xi)$ and $G(\xi)$, respectively. As the fiber path in the squeezing regions shows some random characteristics, it is more reasonable if the amplitude is a random magnitude.

5.2.2 FIBER VOLUME FRACTION

For $m \times n$ rectangle braided composites, where m and n are the braiding yarn numbers along two directions in the cross section of the braided composites, the volume fractions of interior, surface, and corner RUCs corresponding to the whole braided structures can be calculated as follows.

When m and n are even numbers,

$$V_i = \frac{2(m-1)(n-1)+2}{2mn+m+n}$$

$$V_s = \frac{3(m+n-4)}{2mn+m+n} \tag{5.10}$$

$$V_c = \frac{8}{2mn+m+n}$$

where V_i, V_s, and V_c are the volume fractions of the interior, surface, and corner RUCs, respectively.

When one of m and n is odd,

$$V_i = \frac{2(m-1)(n-1)}{2mn+m+n}$$

$$V_s = \frac{3(m+n-2)}{2mn+m+n}$$ (5.11)

$$V_c = \frac{4}{2mn+m+n}$$

If the magnitude of m or n is small, the volume fraction of surface RUC can reach 40%. Therefore, the influence of surface braid configuration cannot be neglected for small specimens.

It is known that a braiding yarn is composed of thousands of fibers. Thus, the fiber volume within the braiding yarn can be represented as

$$\varepsilon = (\pi D_y^2)/4ab$$ (5.12)

where the cross section of braiding yarn is assumed to have elliptical shape. D_y is the equivalent diameter of braiding yarn.

The volume of interior RUC can be calculated as

$$U_i = h^3\tan^2\gamma$$ (5.13)

Then, the yarn volume within the interior RUC is

$$Y_i = 8\pi abh/\cos\gamma$$ (5.14)

Combing with Equations 5.12, 5.13, and 5.14, the fiber volume fraction of the interior RUC is

$$V_{if} = \frac{Y_i\varepsilon}{U_i} = \pi\sqrt{3}\varepsilon/8$$ (5.15)

Similarly, the volume of surface RUC can be calculated as

$$U_s = \sqrt{2}h^3\tan^2\gamma/8$$ (5.16)

The volume fraction of braiding yarn within surface RUC can be expressed as

$$Y_s = \frac{\pi abh}{2}\left(\frac{1}{\cos\gamma}+\frac{3}{\cos\theta}\right)$$ (5.17)

The fiber volume fraction of surface RUC can be calculated by using Equations 5.15, 5.16, and 5.17, which can be written as

$$V_{sf} = \frac{Y_s \varepsilon}{U_s} = \frac{\sqrt{6}\varepsilon\pi}{32}\left(1 + \frac{3\cos\gamma}{\cos\theta}\right) \tag{5.18}$$

Again, the volume of corner RUC can be expressed as

$$U_c = \left(1 + \sqrt{2}\right)h^3\tan^2\gamma/32 \tag{5.19}$$

The yarn volume fraction of corner RUC can be obtained by

$$Y_c = 3\pi abh/2\cos\beta \tag{5.20}$$

The fiber volume fraction for corner RUC can be obtained by using Equations 5.18, 5.19, and 5.20, that is,

$$V_{cf} = \frac{Y_c \varepsilon}{U_c} = \frac{3\sqrt{3}\varepsilon\pi\cos\gamma}{4\left(1 + 2\sqrt{2}\right)\cos\beta} \tag{5.21}$$

Finally, the fiber volume fraction of the whole braided composites can be expressed as

$$V_f = V_i V_{if} + V_s V_{sf} + V_c V_{cf} \tag{5.22}$$

5.2.3 Mesh Generation

When the complex mesogeometric features of textile composites are considered in geometry models using FEM, it brings about a problem in generating meshes. There are some methods to generate meshes of braided composites with complex microstructures. The classical and general strategy to generate meshes is to use a finite element preprocessor or some special textile software packages such as WiseTex when the interface between braiding yarn and matrix are perfectly connected (Lomov et al. 2002). But it is unfavorable when the microgeometry is too complex. One way to overcome these difficulties is to use the digital image–based FEM technique. Pixel-based (two-dimensional [2D]) and voxel-based (3D) meshing concepts have been developed and employed to construct the finite element models. This technique based on image processing can produce locally refined meshes automatically to describe complex 3D geometries accurately. A variety of voxel meshing algorithms have been applied to the analysis of textile composites (Kim and Swan 2003). Because a finite element is identified by each pixel or voxel, the finite elements exhibit uniformity. Therefore, it will lead to models with high computational cost. To improve the computation efficiency, adaptive mesh refinement is combined with pixel/voxel meshing. In the aforementioned mesh generation methods, meshes all coincide with the boundary of mesogeometry of yarn. Therefore, it will need a large number of meshes to obtain conformal meshes on the boundary. To overcome these difficulties, the RUC is meshed with the separated domains where the yarns are tied and embedded into matrix using constraint equations (Xu and Waas 2014), such as binary model (Xu et al. 1995; Iarve et al. 2009) and domain superposition technique (DST) (Jiang, Hallett, and Wisnom 2008). One approach, called the mixed finite element method (MFEM), also uses regular meshes. If an element is cut by an interface, the integration points at the two sides of the interface have different materials (Zohdi et al. 1998; Chen, Tao, and Choy1999b; Zeng et al. 2004). This numerical strategy can calculate the homogenized parameters and be extended to handle the progressive damage analysis of the RUC. However, as there are different materials at different integration points within one element, the large variation in gradients of local displacement can result in a slow rate of convergence for the MFEM. Another potential strategy is extended finite element method (XFEM) (Belytschko and Black 1999; Dolbow and Belytschko 1999; Belytschko et al. 2001), which is based on the concept of partition of unity. The meshes are not necessary to match the real physical surfaces in the geometries. Discontinuities can be classified into strong and weak forms. Strong discontinuities are associated with the displacement variables, whereas weak discontinuities are relevant to the gradient of displacement variables. Typical strong and weak discontinuities are cracks and interfaces between different materials, respectively. For textile composites, material interfaces as weak discontinuities are required to be handled by enrichment techniques in XFEM. Belytschko et al. utilized voxel-/pixel-based meshing to combine implicit surface descriptions of engineering components in structured finite element analysis. An array of voxels sometimes became a background mesh. The level set function was calculated on the nodes of elements to determine the location of implicit surface (Belytschko et al. 2003).

As mentioned earlier, besides the traditional mesh generation methods, many advanced techniques (voxel FEM, DST, MFEM, XFEM, etc.) with different mesh schemes are used to generate finite element meshes of braided composites. The braiding yarn and matrix can be identified by each voxel for the voxel-based meshing techniques. It seems to have potential to combine the μCT scanning experimental method and the voxel-based meshing techniques to mesh the complex mesostructures of braided composites.

5.3 MODELING OF ELASTIC PROPERTIES

Since the 1980s, there exist a variety of theoretical models, such as mosaic model, crimp model, Bridge model, fiber inclination model (Ishikawa and Chou 1982; Yang, Ma, and Chou 1986), strain energy model (Ma, Yang, and Chou 1986), fabric geometry model (Pastore and Ko 1990), global/local model (Whitcomb, Srirengan, and Chapman 1995), mixed volume averaging technique (Wang and Wang 1995a,b), three-cell model (Wu 1996), Mori-Tanaka theories combined with stiffness averaging method (Liang, Du, and Han 1997), and so on, to evaluate the mechanical properties of textile composites. The influence of yarn orientation, fiber volume fraction, and properties of constituents on the stiffness of textile composites can be considered efficiently in these models. These theoretical and computational models can be categorized into four types: stiffness average method, inclusion theory, multiscale

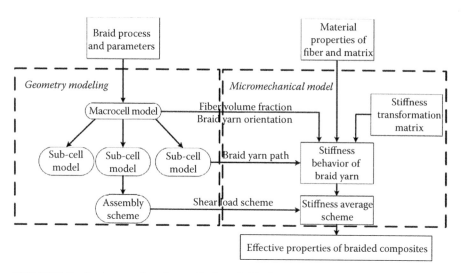

FIGURE 5.6 Basic procedure for prediction of effective properties.

method, and finite element model. The majority of stiffness average methods are mainly based on laminate theory and stiffness transformation to predict the effective behavior of textile composites. The inclusion theory could calculate the microstress of constituents when macroloading is known and could establish the relationships between constituents. The multiscale method tries to establish the relationship of mechanical properties from the microlevel, mesolevel, and macrolevel. The finite element model could consider the realistic textile configuration and calculate the stress and strain fields of constituents, which are also usually utilized to complement the multiscale computation.

For the effective analysis of braided composite properties, geometry modeling and micromechanical analysis should be conducted, as shown in Figure 5.6. Geometry modeling, as mentioned in Section 5.2, is relevant to fiber volume fraction, braiding yarn path and interactive effects among braiding yarns in different orientation. The micromechanical analysis is performed to calculate the stiffness properties of constituents and to determine the shear load scheme among the subcells.

5.3.1 EFFECTIVE PROPERTIES OF BRAIDING YARN

Generally, the alignment of braiding yarn in a braided composite can be assumed to be similar to that in unidirectional fiber reinforced composites. Therefore, the effective properties of braiding yarn can be calculated by the theoretical model used for unidirectional fiber reinforced composites. The longitudinal elastic modulus, E_{11}, and Poisson's ratio, v_{12}, can be obtained by the rule of mixture:

$$E_{11} = V_f E_{11f} + (1 - V_f) E_m \tag{5.23}$$

$$v_{12} = V_f v_{12f} + (1 - V_f) v_m \tag{5.24}$$

where E_{11f} and E_m are the fiber longitudinal modulus and the matrix modulus, respectively; v_{12f} and v_m are the fiber longitudinal Poisson's ratio and matrix Poisson's ratio, respectively; and V_f is fiber volume fraction of braiding yarn.

The transverse modulus, E_{22}, of unidirectional fiber reinforced composites can be expressed explicitly as follows by using Aboudi square cell model (Morais 2000):

$$E_{22} = \frac{\sqrt{V_f}}{\dfrac{\sqrt{V_f}}{E_{22f}} + (1 - \sqrt{V_f})\dfrac{1 - v_m^2}{E_m}} + (1 - \sqrt{V_f})\frac{E_m}{1 - v_m^2} \tag{5.25}$$

where E_{22f} is the fiber transverse modulus.

The volumetric modulus of unidirectional composites in the plane strain can be calculated using the self-consistent theory, that is,

$$k_2 = \frac{(k_{f2} + G_m)k_m + V_f G_m (k_{f2} - G_m)}{k_{f2} + G_m - V_f (k_{f2} - G_m)} \tag{5.26}$$

where k_{f2} and k_m are the composite's fiber transverse and matrix volumetric moduli, respectively, and G_m is the matrix shear modulus.

Using the Equations 5.24 through 5.26, the transverse Poisson's ratio of unidirectional composites can be expressed as

$$v_{23} = 1 - E_{22}\left[(v_{12}^2)/E_{11} + 1/(2k_2)\right] \tag{5.27}$$

To obtain precise prediction of effective properties of unidirectional composites, empirical and semiempirical methods are often used in engineering. Chamis provides an empirical formula to predict the effective properties of unidirectional composites (Li and Wongsto 2004). Owing to the better predicted results, this empirical method is suggested by many scholars to predict the effective properties of unidirectional composites. In this method, longitudinal elastic modulus and Poisson's ratio have the same expressions as Equations 5.23 and 5.24. Transverse elastic modulus, E_{22}, longitudinal shear modulus, G_{12}, and transverse shear modulus, G_{23}, have the following expressions:

$$E_{22} = E_{33} = \frac{E_m}{1 - \sqrt{V_f}\left(1 - \dfrac{E_m}{E_{22f}}\right)} \tag{5.28}$$

$$G_{12} = G_{13} = \frac{G_m}{1 - \sqrt{V_f}\left(1 - \dfrac{G_m}{G_{12f}}\right)} \tag{5.29}$$

$$G_{23} = \cfrac{G_m}{1 - \sqrt{V_f}\left(1 - \cfrac{G_m}{G_{23f}}\right)} \qquad (5.30)$$

Halpin and Tsai also proposed a simple and commonly adopted semiempirical formula (Halpin 1992; Wang 1999), which can be expressed as

$$\frac{M}{M_m} = \frac{1 + \xi \Phi V_f}{1 - \Phi V_f} \qquad (5.31)$$

$$\Phi = \cfrac{\cfrac{M_f}{M_m} - 1}{\cfrac{M_f}{M_m} + \xi} \qquad (5.32)$$

where M is the elastic constant of fiber reinforced composites, such as E_{11}, E_{22}, E_{33}, and v_{12}. M_f is the elastic constant of fiber, such as E_f, G_f, and v_f. M_m is the elastic constant of matrix, such as E_m, G_m, and v_m. ξ is the parameter for measuring the reinforcement efficiency of fiber, which varies from 0 to ∞. The value of ξ is determined from fiber geometry, fiber arrangement, and loading conditions.

When $\xi = 0$, Equations 5.31 and 5.32 become

$$\frac{1}{M} = \frac{V_f}{M_f} + \frac{1 - V_f}{M_m} \qquad (5.33)$$

When $\xi = \infty$, Equations 5.31 and 5.32 can be simplified as

$$M = M_f V_f + (1 - V_f) M_m \qquad (5.34)$$

It can be found that Equations 5.33 and 5.34 are the lower and upper limits for elastic properties of fiber reinforced composites. With the increase of ξ, the effect of fiber reinforcement becomes higher. In general, ξ is equal to 0 when longitudinal elastic modulus and Poisson's ratio are calculated. Value of ξ is set to 2 and 1 for transverse modulus and shear modulus, respectively. In comparison with the experimental results, the Halpin–Tsai semiempirical formula can obtain good prediction when the fiber volume fraction is not very close to 1.

5.3.2 FINITE ELEMENT METHOD

The elastic properties of braided composites can be obtained using RUC with FEM. Periodic boundary conditions should be imposed on the RUC. Usually, the macroscopic stress and strain of braided composites can be expressed as

$$\{\sigma^M\} = [C^M]\{\{\varepsilon^M\} - \{\alpha^M\}\Delta T\} \qquad (5.35)$$

That is,

$$
\begin{Bmatrix} \sigma_1^M \\ \sigma_2^M \\ \sigma_3^M \\ \sigma_4^M \\ \sigma_5^M \\ \sigma_6^M \end{Bmatrix} = \begin{bmatrix} C_{11}^M & C_{12}^M & C_{13}^M & C_{14}^M & C_{15}^M & C_{16}^M \\ C_{21}^M & C_{22}^M & C_{23}^M & C_{24}^M & C_{25}^M & C_{26}^M \\ C_{31}^M & C_{32}^M & C_{33}^M & C_{34}^M & C_{35}^M & C_{36}^M \\ C_{41}^M & C_{42}^M & C_{43}^M & C_{44}^M & C_{45}^M & C_{46}^M \\ C_{51}^M & C_{52}^M & C_{53}^M & C_{54}^M & C_{55}^M & C_{56}^M \\ C_{61}^M & C_{62}^M & C_{63}^M & C_{64}^M & C_{65}^M & C_{66}^M \end{bmatrix} \left(\begin{Bmatrix} \varepsilon_1^M \\ \varepsilon_2^M \\ \varepsilon_3^M \\ \varepsilon_4^M \\ \varepsilon_5^M \\ \varepsilon_6^M \end{Bmatrix} - \begin{Bmatrix} \alpha_1^M \\ \alpha_2^M \\ \alpha_3^M \\ \alpha_4^M \\ \alpha_5^M \\ \alpha_6^M \end{Bmatrix} \Delta T \right) \tag{5.36}
$$

where $\{\sigma^M\}$ and $\{\varepsilon^M\}$ are the macroscopic stress and strain, $\{\alpha^M\}$ is the macroscopic thermal expansion coefficient, $[C^M]$ is the stiffness matrix, and ΔT is the temperature difference.

It can be found from Equation 5.36 that $[C^M]$ can be obtained if 36 independent equations are constructed and the macroscopic stress $\{\sigma^M\}$ and strain $\{\varepsilon^M\}$ or thermal strain $\{\alpha^M\}\nabla T$ are known in advance. If the mechanical strain is applied on the RUC, and the thermal strain does not appear, the macrostress and macrostrain curve can be expressed as

$$
\{\sigma^M\} = [C^M]\{\varepsilon^M\} \tag{5.37}
$$

The macrodisplacements are imposed on the RUC in 1, 2, and 3 normal directions and 12, 13, and 23 shear directions, respectively. The boundary conditions of RUC are applied six times. For example, when a nonzero displacement is applied to the 1 direction and the displacements in other directions are all zero, the first column of matrix can be obtained after the macrostress of the RUC is calculated.

If the thermal strain is applied to the RUC and the mechanical strain is set to zero, the macrostress and macrostrain relation can be rewritten as

$$
\{\sigma^M\} = [C^M]\{-\{\alpha^M\}\Delta T\} \tag{5.38}
$$

The purpose of the aforementioned relation is to produce a nonzero macrostrain and other five zero macrostrains of RUC by applying a temperature difference. Therefore, the properties of constituents within the RUC should be transformed into the global coordinate. The stiffness matrix of braiding yarn in the global coordinate can be expressed as

$$
[\bar{C}] = [T_\sigma][\tilde{C}][T_\sigma]^T \tag{5.39}
$$

where $[\tilde{C}]$ is the stiffness matrix of braiding yarn in the local coordinate along the fiber direction; $[\bar{C}]$ is the stiffness matrix of braiding yarn in the global coordinate; and $[T_\sigma]$ is the stress transformation matrix, which can be written as

$$[T_\sigma] = \begin{bmatrix} l_1^2 & m_1^2 & n_1^2 & 2m_1n_1 & 2n_1l_1 & 2l_1m_1 \\ l_2^2 & m_2^2 & n_2^2 & 2m_2n_2 & 2n_2l_2 & 2l_2m_2 \\ l_3^2 & m_3^2 & n_3^2 & 2m_3n_3 & 2n_3l_3 & 2l_3m_3 \\ l_2l_3 & m_2m_3 & n_2n_3 & m_2n_3+n_2m_3 & n_2l_3+l_2n_3 & l_2m_3+m_2l_3 \\ l_3l_1 & m_3m_1 & n_3n_1 & m_3n_1+n_3m_1 & n_3l_1+l_3n_1 & l_3m_1+m_3l_1 \\ l_1l_2 & m_1m_2 & n_1n_2 & m_1n_2+n_1m_2 & n_1l_2+l_1n_2 & l_1m_2+m_1l_2 \end{bmatrix} \quad (5.40)$$

In this computation, the material properties of constituents are all in the global coordinate. The thermal expansion coefficient of constituents should be constructed to obtain a nonzero macrostrain and other five zero macrostrains of RUC by applying a temperature difference. To calculate the macrostiffness matrix of the RUC, computations should be conducted six times by applying six different loadings. The thermal expansion coefficient for the six different loading conditions can be written in a matrix as follows:

$$\tilde{\alpha} = \begin{bmatrix} -1 & 0 & 0 & 0 & 0 & 0 \\ 0 & -1 & 0 & 0 & 0 & 0 \\ 0 & 0 & -1 & 0 & 0 & 0 \\ 0 & 0 & 0 & -1 & 0 & 0 \\ 0 & 0 & 0 & 0 & -1 & 0 \\ 0 & 0 & 0 & 0 & 0 & -1 \end{bmatrix} \quad (5.41)$$

where $\tilde{\alpha}$ is the thermal expansion coefficient of constituents within the RUC. In the commercial finite element software ABAQUS, the different thermal expansion coefficients can be complemented by the user subroutine UEXPAN () and a sentence "*Expansion, type = ANISO, user" should be added in the .inp file of ABAQUS.

The volume average for macrostress of RUC can be expressed as

$$\sigma_i^M = \frac{\sum_{j=1}^{n_e}(\sigma_i^j V_j)}{\sum_{j=1}^{n_e}(V_j)} \quad (i = 1, 2, 3, 4, 5, 6) \quad (5.42)$$

where V_j is the element volume, σ_i^j is the stress component of element j in the i direction, and n_e is the total element number of the RUC.

Similarly, the macrostrain of RUC can be expressed as

$$\varepsilon_i^M = \frac{\sum_{j=1}^{n_e}(\varepsilon_i^j V_j)}{\sum_{j=1}^{n_e}(V_j)} \quad (i = 1, 2, 3, 4, 5, 6) \quad (5.43)$$

5.3.3 PERIODIC BOUNDARY CONDITIONS

To keep force continuity and displacement compatibility in the opposite faces of a RUC, periodic boundary conditions should be imposed in the simulation.

The expressions for periodic boundary conditions for RUC have been provided in the literature (Whitcomb, Chapman, and Tang 2000; Xia, Zhang, and Ellyin 2003; Li and Wongsto 2004; Li 2008). For the RUC with periodic characteristics, the periodic displacement field can be expressed as

$$u_i = \bar{\varepsilon}_{ik} x_k + u_i^* \tag{5.44}$$

where $\bar{\varepsilon}_{ik}$ is the average strain of the RUC; x_k ($k = 1,2,3$) is the coordinate of any point in the RUC; and u_i^* is the periodic displacement correction value, which is dependent on the loading conditions.

It should be noted that Equation 5.44 is difficult to be applied to the periodic structures in the finite element analysis. On the parallel surface of RUC for textile composites, the periodic displacement field can be expressed as

$$u_i^{j+} = \bar{\varepsilon}_{ik} x_k^{j+} + u_i^* \tag{5.45}$$

$$u_i^{j-} = \bar{\varepsilon}_{ik} x_k^{j-} + u_i^* \tag{5.46}$$

where the superscripts $j+$ and $j-$ represent the two parallel surfaces.

It is known that u_i^* is the same on the parallel surfaces. Using Equations 5.45 and 5.46, an equation can be established as follows:

$$u_i^{j+} - u_i^{j-} = \bar{\varepsilon}_{ik}(x_k^{j+} - x_k^{j-}) = \bar{\varepsilon}_{ik} \Delta x_k^j \tag{5.47}$$

For each group of parallel surfaces, Δx_k^j is a constant value. If $\bar{\varepsilon}_{ik}$ is provided, the right side of Equation 5.47 can be rewritten as

$$u_i^{j+}(x, y, z) - u_i^{j-}(x, y, z) = c_i^j \quad (i, j = 1, 2, 3) \tag{5.48}$$

The aforementioned equation does not have the correction variable u_i^*. Therefore, it is easy to be carried out by using multipoint constraint equation in finite element analysis.

In displacement finite element analysis, stress boundary conditions belong to natural boundary conditions. Thus, the periodic displacement condition applied to RUC can ensure the results' uniqueness. In the finite element model, linear constraint equations are imposed on the node pairs corresponding to the parallel surfaces. For a hexahedron RUC, as shown in Figure 5.7, the length, width, and height are expressed as W_x, W_y, and h, respectively. It can be found that the node pairs on the parallel surface should satisfy one-directional constraint equations. But the node pairs on the edges or corner should satisfy two- or three-directional constraint equations at the same time. These constraint equations may not be independent, and this may result in difficult convergence in the finite element analysis. Therefore, some constraint

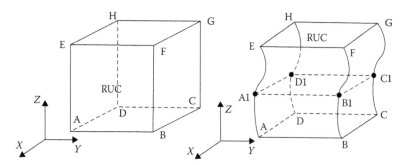

FIGURE 5.7 Hexahedral representative unit cell for applying periodic boundary conditions.

equations should be simplified into independent equations for the node pairs on the edges or on the corner. The detailed constraint equations applied to the node pairs located on the corresponding surfaces, edges, and corner are provided as follows.

If the node pairs are on the surfaces with x normal direction, the constraint equations can be written as

$$\begin{cases} u\big|_{x=W_x} - u\big|_{x=0} = W_x \varepsilon_x^0 \\ v\big|_{x=W_x} - v\big|_{x=0} = 0 \\ w\big|_{x=W_x} - w\big|_{x=0} = 0 \end{cases} \tag{5.49}$$

For the node pairs on the surfaces with y normal direction, similarly,

$$\begin{cases} u\big|_{y=W_y} - u\big|_{y=0} = W_y \gamma_{xy}^0 \\ v\big|_{y=W_y} - v\big|_{y=0} = W_y \varepsilon_y^0 \\ w\big|_{y=W_y} - w\big|_{y=0} = 0 \end{cases} \tag{5.50}$$

For the node pairs on the surfaces with z normal direction,

$$\begin{cases} u\big|_{z=h} - u\big|_{z=0} = h\gamma_{xz}^0 \\ v\big|_{z=h} - v\big|_{z=0} = h\gamma_{yz}^0 \\ w\big|_{z=h} - w\big|_{z=0} = h\varepsilon_z^0 \end{cases} \tag{5.51}$$

where ε_x^0, ε_y^0, ε_z^0, γ_{xy}^0, γ_{xz}^0, and γ_{yz}^0 are the macrostrains of the RUC.

There are 12 edges in the hexahedron RUC, as shown in Figure 5.7, which can be divided into three groups. The edges AD, BC, FG, and FH are all parallel to the x axis. The edges CD, BA, FE, and GH are parallel to the y axis. And the edges HD,

EA, FB, and GC are parallel to the z axis. Taking the group that is parallel to the z axis, for example, three linear constraint equations can be written as follows, where HD is the reference edge:

$$\begin{cases} u_{EA} - u_{HD} = W_x\varepsilon_x^0 \\ v_{EA} - v_{HD} = 0 \\ w_{EA} - w_{HD} = 0 \end{cases}, \begin{cases} u_{FB} - u_{HD} = W_x\varepsilon_x^0 + W_y\gamma_{xy}^0 \\ v_{FB} - v_{HD} = W_y\varepsilon_y^0 \\ w_{FB} - w_{HD} = 0 \end{cases}, \text{and} \begin{cases} u_{GC} - u_{HD} = W_y\gamma_{xy}^0 \\ v_{GC} - v_{HD} = W_y\varepsilon_y^0 \\ w_{GC} - w_{HD} = 0 \end{cases} \quad (5.52)$$

The periodic conditions of the node pairs in the other two groups can be written with similar forms as in Equation 5.52.

There are eight corner points in the RUC. The constraint equations for the corner nodes should also be simplified. The corner node D is considered as the reference point, and the constraint equations for corner nodes E, F, and D can be written in the following forms.

$$\begin{cases} u_E - u_D = W_x\varepsilon_x^0 + h\gamma_{xz}^0 \\ v_E - v_D = h\gamma_{yz}^0 \\ w_E - w_D = h\varepsilon_z^0 \end{cases} \quad (5.53)$$

$$\begin{cases} u_F - u_D = W_x\varepsilon_x^0 + W_y\gamma_{xy}^0 + h\gamma_{xz}^0 \\ v_F - v_D = W_y\varepsilon_y^0 + h\gamma_{yz}^0 \\ w_F - w_D = h\varepsilon_z^0 \end{cases} \quad (5.54)$$

$$\begin{cases} u_G - u_D = W_y\gamma_{xy}^0 + h\gamma_{xz}^0 \\ v_G - v_D = W_y\varepsilon_y^0 + h\gamma_{yz}^0 \\ w_G - w_D = h\varepsilon_z^0 \end{cases} \quad (5.55)$$

The constraint equations for the other corner nodes A, B, C, and H with the reference point D can be obtained easily. The aforementioned periodic boundary conditions are mainly used for applying mechanical strain on the RUC. For imposing thermal strain on the RUC, the periodic boundary can be simplified considerably. The parallel surfaces within the RUC can be constrained to satisfy the periodic boundary conditions using the "*Tie" command in ABAQUS.

5.3.4 HOMOGENIZATION METHOD

5.3.4.1 Basic Theories

Homogenization theory as a strict mathematical method was developed in the 1970s (Babuška 1976). Some scholars have used two-scale asymptotic analysis method to study the mechanical properties of composites. Asymptotic technique can transform a problem into a mesoscale problem and a macrohomogenization problem. The macroscopic mechanical properties of the whole structure can be obtained by using a mesoscale RUC or representative volume element (RVE). In this theory, two scales are introduced: a macroscale (global) coordinate x and a mesoscale (local) coordinate y. Then, a small parameter $\varepsilon\,(0 < \varepsilon \ll 1)$ is introduced, which is the ratio of realistic length of unit vector in mesoscale coordinate to realistic length of unit vector in macroscale coordinate. It satisfies $y + nY = x/\varepsilon$, where Y is the size of RUC or RVE and n is equal to 1, 2, 3, ...

Owing to the nonhomogeneous nature of composites, the interior temperature, displacement, and stress of composites at the small region around the macroposition x have great variations. These physical functions are dependent on not only the macrocoordinate x but also the mesocoordinate y. Therefore, the function $\phi^\varepsilon(x)$ representing the interior temperature, displacement, and stress should also be related to the coordinate y, that is, $\phi^\varepsilon(x) = \phi(x, y)$. Owing to the periodic characteristics of mesostructures of the material, the function dependent on the mesocoordinate should also keep the periodic characteristics, which can be expressed as

$$\phi^\varepsilon(x) = \phi(x, y) = \phi(x, y + Y) \tag{5.56}$$

The displacement can be expanded asymptotically by the small parameter ϵ, that is,

$$u_i = u_i^\varepsilon(x) = u_i(x, y) = u_i^0(x, y) + \epsilon u_i^1(x, y) + \epsilon^2 u_i^2(x, y) + \dots \tag{5.57}$$

The strain ε_{ij} can be calculated from Equation 5.57 as follows:

$$\varepsilon_{ij}(u^\varepsilon) = \frac{1}{2}\left(\frac{\partial u_i^\varepsilon}{\partial x_j^\varepsilon} + \frac{\partial u_j^\varepsilon}{\partial x_i^\varepsilon}\right) = \frac{1}{\epsilon}\varepsilon_{ij}^{(-1)}(x, y) + \varepsilon_{ij}^0(x, y) + \epsilon\varepsilon_{ij}^{(1)}(x, y) + \epsilon^2 \dots \tag{5.58}$$

The stiffness tensor, C_{ijkl}^ε, of material should have the following relation due to the periodic characteristic of the material:

$$C_{ijkl}^\varepsilon(x) = C_{ijkl}^\varepsilon(x/\epsilon) = C_{ijkl}^\varepsilon(y) \tag{5.59}$$

Therefore, the stress can be calculated by using the constitutive relation, that is,

$$\begin{aligned}
\sigma_{ij}^\varepsilon &= C_{ijkl}^\varepsilon \varepsilon_{kl} \\
&= \frac{1}{\epsilon}C_{ijkl}^\varepsilon \varepsilon_{ij}^{(-1)}(x, y) + C_{ijkl}^\varepsilon \varepsilon_{ij}^{(0)}(x, y) + \epsilon C_{ijkl}^\varepsilon \varepsilon_{ij}^{(1)}(x, y) + \dots \\
&= \frac{1}{\epsilon}\sigma_{ij}^{(-1)}(x, y) + \sigma_{ij}^{(0)}(x, y) + \epsilon\sigma_{ij}^{(1)}(x, y) + \dots
\end{aligned} \tag{5.60}$$

Then, Equation 5.60 is substituted into the stress equilibrium equation $\sigma_{ij,j} + f_i = 0$, which can be written in the ϵ power form:

$$\begin{aligned}
&\epsilon^{-2}\frac{\partial \sigma_{ij}^{(-1)}}{\partial y_j} + \epsilon^{-1}\left(\frac{\partial \sigma_{ij}^{(-1)}}{\partial x_j} + \frac{\partial \sigma_{ij}^{(0)}}{\partial y_j}\right) + \epsilon^0\left(\frac{\partial \sigma_{ij}^{(0)}}{\partial x_j} + \frac{\partial \sigma_{ij}^{(1)}}{\partial y_j} + f_i\right) \\
&+ \epsilon^1\left(\frac{\partial \sigma_{ij}^{(1)}}{\partial x_j} + \frac{\partial \sigma_{ij}^{(2)}}{\partial y_j}\right) + \epsilon^2 \dots = 0
\end{aligned} \tag{5.61}$$

When $\epsilon \to 0$, the coefficient of Equation 5.65 should be equal to zero. Combining with Equations 5.57, 5.58, and 5.60, a series of asymptotic equations can be obtained:

$$\epsilon^{-2}: \frac{\partial}{\partial y_j} C_{ijkl} \frac{\partial u_k^{(0)}}{\partial y_l} = 0 \tag{5.62a}$$

$$\epsilon^{-1}: \frac{\partial}{\partial x_j} C_{ijkl} \frac{\partial u_k^{(0)}}{\partial y_l} + \frac{\partial}{\partial y_j} C_{ijkl} \left(\frac{\partial u_k^{(0)}}{\partial y_l} + \frac{\partial u_k^{(1)}}{\partial y_l} \right) = 0 \tag{5.62b}$$

$$\epsilon^0: \frac{\partial}{\partial x_j} C_{ijkl} \left(\frac{\partial u_k^{(0)}}{\partial y_l} + \frac{\partial u_k^{(1)}}{\partial y_l} \right) + \frac{\partial}{\partial y_j} C_{ijkl} \left(\frac{\partial u_k^{(1)}}{\partial y_l} + \frac{\partial u_k^{(2)}}{\partial y_l} \right) + f_i = 0 \tag{5.62c}$$

$$\epsilon^1: \frac{\partial}{\partial x_j} C_{ijkl} \left(\frac{\partial u_k^{(1)}}{\partial y_l} + \frac{\partial u_k^{(2)}}{\partial y_l} \right) + \frac{\partial}{\partial y_j} C_{ijkl} \left(\frac{\partial u_k^{(2)}}{\partial y_l} + \frac{\partial u_k^{(3)}}{\partial y_l} \right) = 0 \tag{5.62d}$$

There exists a theorem for the Y periodic function, $\phi(y)$, which is

$$\lim_{\epsilon \to 0^+} \int_{\Omega^\epsilon} \phi(x/\epsilon) \, dV = \frac{1}{|Y|} \int_\Omega \left(\int_Y \phi(y) \, dY \right) dV \tag{5.63}$$

where $|Y|$ is the volume of RUC.

Multiplying $\delta u_i^{(0)}$, integrating on the region, and solving the limit for $\epsilon \to 0^+$, Equation 5.62a can be rewritten using Equation 5.63:

$$\lim_{\epsilon \to 0^+} \int_{\Omega^\epsilon} \frac{\partial}{\partial y_j} \left(C_{ijkl}^\epsilon \frac{\partial u_k^{(0)}}{\partial y_l} \right) \delta u_i^{(0)} \, dV = \frac{1}{|Y|} \int_\Omega \int_Y \frac{\partial}{\partial y_j} \left(C_{ijkl} \frac{\partial u_k^{(0)}}{\partial y_l} \right) \delta u_i^{(0)} \, dY \, dV \tag{5.64}$$

Then, using partial integration, Gauss theorem, and $\delta u_i^{(0)} = 0$ on the boundary, we get $S \, (S = \partial Y)$.

Equation 5.64 can result in

$$\frac{\partial u_k^{(0)}}{\partial y_l} = 0 \tag{5.65}$$

It can be found that $u_k^{(0)}$ is only the function of macrocoordinate x. Equation 5.58 can be rewritten as

$$u_i^\epsilon(x) = u_i^{(0)}(x) + \epsilon u_i^{(1)}(x,y) + \epsilon^2 u_i^{(2)}(x,y) + \ldots \tag{5.66}$$

where $u_i^{(0)}$ can be recognized as the macrodisplacement and $u_i^{(1)}$ and $u_i^{(2)}$ are the mesodisplacements. It can be explained physically that there is a realistic perturbation displacement, u_i^ϵ, around the macrodisplacement $u_i^{(0)}$, where $u_i^{(1)}, u_i^{(2)}, \ldots$ and ϵ are used to describe the small perturbation displacement.

Substituting Equation 5.65 in Equation 5.62b, the one-order mesodisplacement can be expressed as

$$\frac{\partial}{\partial y_j}\left(C_{ijkl}\frac{\partial u_k^{(0)}}{\partial y_l}\right) = -\frac{\partial C_{ijkl}}{\partial y_j}\frac{\partial u_k^{(1)}}{\partial y_l} \tag{5.67}$$

Equation 5.67 can be solved by FEM. A characteristic function, $(\chi_k^{mn}(y_i))$, can establish the relation between the macrodisplacement and the one-order displacement:

$$u_i^{(1)} = \chi_i^{kl}\frac{\partial u_k^{(0)}}{\partial x_l} \tag{5.68}$$

It can be demonstrated that the characteristic function should satisfy

$$\frac{\partial}{\partial y_j}\left(C_{ijkl}\frac{\partial \chi_k^{mn}}{\partial y_l}\right) = -\frac{\partial C_{ijkl}}{\partial y_j} \tag{5.69}$$

Subsituting Equation 5.68 into Equation 5.58, $\varepsilon_{ij}^{(0)}$ and $\sigma_{ij}^{(0)}$ can be written as

$$\varepsilon_{ij}^{(0)}(x,y) = \left(\delta_{im}\delta_{jn} + \frac{\partial \chi_i^{mn}}{\partial y_j}\right)\frac{\partial u_m^{(0)}}{\partial x_n} \tag{5.70}$$

$$\sigma_{ij}^{(0)}(x,y) = C_{ijkl}(y)\varepsilon_{ij}^{(0)}(x,y) = C_{ijkl}(y)\left(\delta_{im}\delta_{jn} + \frac{\partial \chi_i^{mn}}{\partial y_j}\right)\frac{\partial u_m^{(0)}}{\partial x_n} \tag{5.71}$$

Similarly, Equation 5.62c can be rewritten as follows by substituting Equation 5.68:

$$\frac{\partial}{\partial x_j}C_{ijkl}\left(\frac{\partial u_k^{(0)}}{\partial x_l} + \frac{\partial \chi_k^{mn}}{\partial y_l}\frac{\partial u_m^{(0)}}{\partial x_n}\right) + \frac{\partial}{\partial y_j}C_{ijkl}\left(\frac{\partial}{\partial x_l}\left(\chi_k^{mn}\frac{\partial u_m^{(0)}}{\partial x_n}\right) + \frac{\partial u_k^{(2)}}{\partial y_l}\right) + f_i = 0 \tag{5.72}$$

where $u_i^{(2)}$ is a Y periodic function. If $u_i^{(2)}$ has a unique solution, it should satisfy the following relationship:

$$\int_Y \frac{\partial}{\partial y_j}C_{ijkl}\frac{\partial u_k^{(2)}}{\partial y_l}dY = 0 \tag{5.73}$$

Substituting Equation 5.72 in Equation 5.73 and dividing $|Y|$ on both sides of the equation, it can be written as

$$\frac{\partial}{\partial x_j}\left\{\frac{1}{|Y|}\int_Y C_{ijkl}\left(\delta_{km}\delta_{ln} + \frac{\partial \chi_k^{mn}}{\partial y_l}\right)dY\frac{\partial u_m^{(0)}}{\partial x_n}\right\} + f_i = 0 \tag{5.74}$$

From Equation 5.74, the homogenization coefficient, C_{ijmn}^H, on the unit cell can be defined as

$$C_{ijmn}^H = \frac{1}{|Y|}\int_Y C_{ijkl}\left(\delta_{km}\delta_{ln} + \frac{\partial \chi_k^{mn}}{\partial y_l}\right)dY \tag{5.75}$$

It can be found that the homogenization coefficient is related to the effective properties of the mesostructure. It is noted that $\chi_k^{mn}(y)$ is only relevent to the mesostructure and constituents of unit cell. After $\chi_k^{mn}(y)$ is obtained, the homogenization coefficient can be calculated using Equation 5.75. The macrodisplacement can be obtained using Equation 5.74. The first-order mesodisplacement, strain, and stress can be calculated using Equations 5.68, 5.70, and 5.71.

5.3.4.2 Thermal Stress Method

Considering the symmetry of the elastic stiffness tenor for subscript kl, Equation 5.67 can be rewritten as

$$\frac{\partial}{\partial y_j}\left(C_{ijkl}\frac{1}{2}\left(\frac{\partial \chi_k^{mn}}{\partial y_l}+\frac{\partial \chi_l^{mn}}{\partial y_k}\right)\right)=-\frac{\partial C_{ijkl}}{\partial y_j} \tag{5.76}$$

In the aforementioned equation, the characteristic function $\chi_k^{mn}(y)$ can be recognized as the thermal deformation due to temperature change (Yuan and Fish 2008). Equation 5.76 can be rewritten as

$$\frac{\partial}{\partial y_j}\left(C_{ijkl}\left[\frac{1}{2}\left(\frac{\partial \chi_k^{mn}}{\partial y_l}+\frac{\partial \chi_l^{mn}}{\partial y_k}\right)+I_{klmn}\right]\right)=0 \tag{5.77}$$

where $I_{klmn}=(\delta_{mk}\delta_{nl}+\delta_{nk}\delta_{ml})/2$. Let $I_{klmn}=-\alpha_{kl}^{mn}\Delta T$; then, Equation 5.75 can be rewritten as

$$C_{ijmn}^H=\frac{1}{|Y|}\int_Y C_{ijkl}\left[\frac{1}{2}\left(\frac{\partial \chi_k^{mn}}{\partial y_l}+\frac{\partial \chi_l^{mn}}{\partial y_k}\right)-\alpha_{kl}^{mn}\Delta T\right]dY=\frac{1}{|Y|}\int_Y \sigma_{ij}^{(mn)}\,dY \tag{5.78}$$

The aforementioned homogenization equation has been converted into the thermal stress equation. If $\Delta T=1$, and α_{kl}^{mn} is chosen to have the form of Equation 5.41, the stress in Equation 5.78 can be calculated by FEM. Then the homogenization coefficient can be obtained.

5.3.4.3 Boundary Force Method

The weak form of Equation 5.77 can be expressed as

$$\int_Y \frac{\partial}{\partial y_j}\left\{C_{ijkl}\left[\frac{1}{2}\left(\frac{\partial \chi_k^{mn}}{\partial y_l}+\frac{\partial \chi_l^{mn}}{\partial y_k}\right)+I_{klmn}\right]\right\}\delta v_i\,dV=0 \tag{5.79}$$

Using Gauss theorem, the first part of Equation 5.79 can be written as

$$\int_Y \frac{\partial}{\partial y_j}\left\{C_{ijkl}\left[\frac{1}{2}\left(\frac{\partial \chi_k^{mn}}{\partial y_l}+\frac{\partial \chi_l^{mn}}{\partial y_k}\right)\right]\right\}\delta v_i\,dV$$
$$=\int_Y C_{ijkl}\frac{1}{2}\left(\frac{\partial \chi_k^{mn}}{\partial y_l}+\frac{\partial \chi_l^{mn}}{\partial y_k}\right)\delta v_i n_j\,dS-\int_Y C_{ijkl}\frac{1}{2}\left(\frac{\partial \chi_k^{mn}}{\partial y_l}+\frac{\partial \chi_l^{mn}}{\partial y_k}\right)\frac{\partial \delta v_i}{\partial y_j}dV \tag{5.80}$$

where n_j is the component of unit normal vector in j direction of boundary ∂Y. The first component of the right hand side of Equation 5.80 is zero because $\delta v_i = 0$ on the boundary ∂Y. And considering the symmetry of C_{ijkl}, the second component of the right hand side of Equation 5.80 can be written as

$$\int_Y C_{ijkl} \frac{1}{2} \left(\frac{\partial \chi_k^{mn}}{\partial y_l} + \frac{\partial \chi_l^{mn}}{\partial y_k} \right) \frac{\partial \delta v_i}{\partial y_j} \, dV$$

$$= \int_Y C_{ijkl} \frac{1}{2} \left(\frac{\partial \chi_k^{mn}}{\partial y_l} + \frac{\partial \chi_l^{mn}}{\partial y_k} \right) \frac{1}{2} \left(\frac{\partial \delta v_i}{\partial y_j} + \frac{\partial \delta v_j}{\partial y_i} \right) dV$$

(5.81)

Using Gauss theorem, the second part of Equation 5.79 can be written as

$$\int_Y \frac{\partial}{\partial y_j} (C_{ijkl} I_{klmn}) \delta v_i \, dV = \int_{\partial Y} C_{ijkl} I_{klmn} \delta v_i n_j \, dS - \int_Y C_{ijkl} I_{klmn} \frac{\partial \delta v_i}{\partial y_j} \, dV$$

$$= -\int_Y C_{ijkl} I_{klmn} \frac{\partial \delta v_i}{\partial y_j} \, dV$$

(5.82)

If there are p regions $Y^1, Y^2, \ldots, Y^p$ in a unit cell, where ∂Y is the boundary of unit cell and $\partial Y^1, \partial Y^2, \ldots, \partial Y^p$ are the boundaries of p regions, the right hand side of Equation 5.82 can be expressed as

$$\int_Y C_{ijkl} I_{klmn} \frac{\partial \delta v_i}{\partial y_j} \, dV = \sum_p \int_{Y^p} C_{ijkl} I_{klmn} \frac{\partial \delta v_i}{\partial y_j} \, dV$$

$$= \sum_p \int_{\partial Y^p} C_{ijkl} I_{klmn} n_j \delta v_i \, dS = \sum_p \int_{\partial Y^p} C_{ijkl} n_j \delta v_i \, dS$$

$$= \sum_p \left\{ \int_{\partial Y^p - \partial Y^p \cap \partial Y} C_{ijmn} n_j \delta v_i \, dS + \int_{\partial Y^p \cap \partial Y} C_{ijmn} n_j \delta v_i \, dS \right\}$$

$$= \sum_p \int_{\partial Y^p - \partial Y^p \cap \partial Y} C_{ijmn} n_j \delta v_i \, dS$$

(5.83)

where $\partial Y^p - \partial Y^p \cap \partial Y$ represents the interior boundary of p regions.
Therefore, Equation 5.79 is equivalent to

$$\int_Y C_{ijkl} \frac{1}{2} \left(\frac{\partial \chi_k^{mn}}{\partial y_l} + \frac{\partial \chi_l^{mn}}{\partial y_k} \right) \frac{1}{2} \left(\frac{\partial \delta v_i}{\partial y_j} + \frac{\partial \delta v_j}{\partial y_i} \right) dV = -\sum_p \int_{\partial Y^p - \partial Y^p \cap \partial Y} C_{ijmn} n_j \delta v_i \, dS \quad (5.84)$$

This equation is similar to the weak formulation of the 3D elastic problem. The difference is that there exist boundary forces that are distributed on the interior boundaries of the unit cell. The boundary force is related to the material's properties and boundary shapes.

For isotropic material, $C_{ijmn} = \lambda\delta_{ij}\delta_{mn} + \mu(\delta_{im}\delta_{jn} + \delta_{in}\delta_{jm})$. Let $F_i^{bmn} = -C_{ijmn}n_j$; then, the matrix expression can be written as

$$F^b = -\begin{bmatrix} (\lambda+2\mu)n_1 & \lambda n_1 & \lambda n_1 & \mu n_2 & \mu n_3 & 0 \\ \lambda n_2 & (\lambda+2\mu)n_2 & \lambda n_2 & \mu n_1 & 0 & \mu n_3 \\ \lambda n_3 & \lambda n_3 & (\lambda+2\mu)n_3 & 0 & \mu n_1 & \mu n_2 \end{bmatrix} \quad (5.85)$$

The column of matrix F^b represents the boundary forces in three directions with a variation of mn. Therefore, the stiffness matrix in the local coordinate should be transformed into the global coordinate as expressed in Equation 5.39.

In the commercial finite element software ABAQUS, the variation in boundary distribution force can be carried out by the user subroutine UTRACLOAD (). A sentence "*Dsload, op = new, follower = No" should be added to the ABAQUS input file.

5.3.4.4 Prediction of Effective Properties of Braided Composites by Homogenization Method

The geometry model for braided composites was established in Section 5.2.1. To illustrate the prediction of properties using homogenization method, four-step braided composites have been developed using T300 carbon fiber as the reinforcing fiber, whose elastic properties are listed in Table 5.2 (Fang et al. 2009, 2011). The mechanical properties of braiding yarn can be obtained by the homogenization

TABLE 5.2
Properties of Constituents of 3D Four-Step Braided Composites

Reinforced Fiber		Matrix		
	T700 12K Carbon Fibers	T300 Carbon Fiber	TDE-85	
Longitudinal tensile elastic modulus, E_{f11}	230 GPa	230 GPa	Elastic modulus E_m	3.45 GPa
Longitudinal compressive elastic modulus, E_{f11}	130 GPa		Poisson's ratio v_m	0.35
Transverse elastic moduli, $E_{f22} = E_{f33}$	18.226 GPa	13.8 GPa	Density	1.22 g/cm³
Longitudinal shear moduli, $G_{f12} = G_{f13}$	36.597 GPa	9.0 GPa		
Longitudinal Poisson's ratios $v_{f12} = v_{f13}$	0.27	0.2		
Transverse Poisson's ratio, v_{f23}	0.3	0.25		
Longitudinal tensile strength, F_t	4900 MPa	2350 MPa		
Cross-section area, A	0.44 mm²			
Diameter of fiber, d_f	0.7 μm			
Density	1.80 g/cm³			

TABLE 5.3

Effective Properties of Braiding Yarn Obtained by Homogenization and Semiempirical Method

	E_1(GPa)	E_2(GPa)	G_{12}(GPa)	G_{13}(GPa)	v_{12}	v_{23}
Homogenization results	120.642	8.698	3.687	2.648	0.2604	0.3655
Semiempirical results	120.870	8.915	4.174	3.215	0.2644	0.3863

TABLE 5.4

Effective Properties of Braided Composites Obtained by Homogenization Method

No.	Homogenization Results		Experimental Results	
	E_L(GPa)	v_{zx}	E_L(GPa)	v_{zx}
1	28.9	0.580	30.0	0.606
2	64.2	0.640	66.8	0.648
3	65.7	0.547	67.3	0.568
4	26.1	0.462	27.1	0.479

method, as listed in Table 5.3, which are in good agreement with the predicted results obtained by Equations 5.23, 5.24, 5.27, 5.28, 5.29, and 5.30. Then, the homogenizaiton results of braiding yarn are used to calculate the mechanical properties of four-step braided composites. It can be found from Table 5.4 that the effective properties of braided composites obtianed from the homogenization method are in good agreement with the experimental results.

5.4 PROGRESSIVE DAMAGE ANALYSIS

Braided composites are multiphase heterogeneous anisotropic materials whose damage mechanisms are more complex than those of homogeneous isotropic materials. Usually, the damage mechanisms of fiber reinforced composites can be categorized into matrix cracking, fiber breakage, and interfacial debonding. Usually, macro, meso, and micro methods are combined together to analyze the complex damage states of braided composites. To recognize multiple damage mechanisms well, reliable damage initiation criteria and damage evolution laws are chosen to simulate the progressive damage and failure process of braided composites. Some failure criteria should be developed to be used as the constituents' damage initiation criteria. Besides, the in situ properties of constituents (braiding yarn and matrix) in 3D braided composites should be determined reasonably.

5.4.1 STRENGTH THEORY FOR BRAIDING YARN

The axial tensile statistical strength of unidirectional fiber reinforced composites was studied by Zhu et al. (1989). As the tensile strength of 3D four-directional braided composites is determined by braiding yarn strength, especially for low-angle braided composites, this statistical theory for yarn can be introduced to obtain the strength of braided composites. The statistical theory is expressed simply as follows:

It is assumed that the probability distribution of fiber strength satisfies the Weibull function, that is,

$$F(\sigma) = 1 - \exp(-\alpha\delta\sigma^\beta) \tag{5.86}$$

where α, β, and δ are the Weibull scale parameter, Weibull shape parameter, and ineffective length, respectively. For the braiding yarn within 3D braided composites, it is assumed that β equals 6 (Zuo, Xiao, and Liao 2007) and α can be determined as $\alpha = \sigma_f^{-\beta}$. The term σ_0 is the minimal fiber stress of the braiding yarn, which can be determined from the study by Wang (1999).

A braiding yarn is composed of N fibers. When the tensile load P is applied on the longitudinal direction of the braiding yarn, N_f fibers are broken. When the applied load increases by dP, the average stress will increase by $d\sigma$. Further, assume that dN_f fibers will break due to the load increment. The broken fibers, N_f, should satisfy the following equation and boundary condition:

$$\frac{dN_f}{d\sigma} + \left[(a+1)\frac{dF(\sigma)}{d\sigma} - a\frac{dF(K\sigma)}{d\sigma}\right]N_f = N\frac{dF(\sigma)}{d\sigma}, \quad N_f(0) = 0 \tag{5.87}$$

here $F(K\sigma)$ is the fiber breakage probability, K is the stress concentration factor, and a is the number of fibers adjacent to a broken fiber. If there are four or six fibers adjacent to a broken fiber, K equals 1.146 or 1.104, respectively. If K equals 1.104, the solution of the aforementioned differential equation is

$$N_f = N\exp[(a+1)\exp(-\alpha\delta\sigma^\beta) - a\exp(-\alpha\delta m\sigma^\beta)]T \tag{5.88}$$

where $T = \int_x^1 \exp[-(a+1)X + aX^m]dX$, $X = \exp(-\alpha\sigma\sigma^\beta)$, and $m = K^\beta$.

The ineffective length δ can be obtained by

$$\delta = \frac{d_f}{2}[(K_f^{-1/2} - 1)\frac{E_{f11}}{2G_m}]^{1/2}Ch^{-1}[\frac{1+(1-\varphi)^2}{2(1-\varphi)}] \tag{5.89}$$

where d_f, K_f, and E_{f11} are the diameter, fiber filling coefficient, and longitudinal elastic modulus of fiber, respectively; G_m is the matrix shear modulus; and φ is the fraction of the undisturbed stress value.

The expectation of average fiber failure stress of braiding yarn is expressed as

$$\overline{\sigma}_B = [1 - N_f(\overline{\sigma}_m)/N]\overline{\sigma}_m \tag{5.90}$$

where $\overline{\sigma}_m$ is the maximum stress of unbroken fiber, which can be obtained by

$$\frac{d}{d\sigma}[(1 - N_f/N)\sigma]\Big|_{\sigma = \overline{\sigma}_m} = 0 \tag{5.91}$$

When N is large, the strength distribution of braiding yarn approaches a normal distribution given by Daniels (1945). The density distribution function is

$$g(\sigma) = \frac{1}{\sqrt{2\pi}\psi_B} \exp[-\frac{(\sigma - \overline{\sigma}_B)}{2\psi_B^2}]$$ (5.92)

where ψ_B is standard deviation of the normal distribution, which can be expressed by

$$\psi_B = \overline{\sigma}_m \{F(\overline{\sigma}_m)[1 - F(\overline{\sigma}_m)]\}^{1/2} N^{-1/2}$$ (5.93)

The braiding yarn can be considered as links in a chain that contains n links. The weakest link theory is used to determine the failure of the braiding yarn. The probability density distribution function, $p(\sigma)$, is defined as

$$p(\sigma) = ng(\sigma)[1 - G(\sigma)]^{n-1}$$ (5.94)

where $G(\sigma)$ is the cumulative distribution function of $g(\sigma)$. The expectation of fiber stress at failure of braiding yarn, $\overline{\sigma}_{fB}$, can be obtained by

$$\mathrm{d}\,p(\sigma)/\mathrm{d}\sigma\,|_{(\sigma = \overline{\sigma}_{fB})} = 0$$ (5.95)

Therefore, the longitudinal tensile strength of undulated braiding yarn, X_{yt}, can be obtained from the literature (Hearle, Grosberg, and Backer 1969):

$$X_{yt} \approx K_f \overline{\sigma}_{fB} \cos^2 \theta$$ (5.96)

where θ is the twisting angle of the braiding yarn.

By considering the effect of matrix yield, the longitudinal compressive strength of the braiding yarn can be obtained by the compressive and shear failure models given by Budiansky and Fleck (1993):

$$X_{yc} = \frac{\overline{G}_m}{(1 - V_f)}, \quad \overline{G}_m = \frac{G_m}{[1 + 3G_m(1/E_{ms} - 1/E_m)]}$$ (5.97)

where E_m is the elastic modulus of matrix and E_{ms} denotes the inelastic tangential modulus.

The transverse tensile strength, transverse compressive strength, and shear strength, denoted as Y_{yt}, Y_{yc}, and S, respectively, can be obtained from the mechanical properties of fiber and matrix (Pastore and Ko 1990).

$$Y_{yt} = \frac{1 + K_f(\frac{1}{\eta_y} - 1)}{K_{my}} X_m$$ (5.98)

$$Y_{yc} = 4Y_{yt}$$ (5.99)

$$S = \frac{1 + K_f(\frac{1}{\eta_s} - 1)}{K_{ms}} S_m \qquad (5.100)$$

In the aforementioned equations, X_m is the tensile strength of matrix; η_y and η_s are Cai experience coefficients; K_{my} and K_{ms} are the tensile and shear stress concentration coefficients, respectively, of matrix; S_m is the shear strength of matrix; and η_y and η_s equal 0.5 (Wang 1999).

5.4.2 Damage Theory for Braiding Yarn

A braiding yarn experiences an initial damage and eventual failure with the increase of external loading. To simulate the failure process of braiding yarn, the damage models for different damage modes of braiding yarn should be established reasonably.

5.4.2.1 Damage Initiation Criterion

The failure mechanisms of braiding yarns within braided composites can be categorized into three types: longitudinal (L direction) fiber–matrix shear failure, fiber buckling failure, and transverse (in TZ plane) matrix failure (Fang, Liang, Lu et al. 2011). For the longitudinal compressive failure of unidirectional fiber composites, buckling and kinking of fibers, accompanied by matrix fragmentation, are observed in uniaxial compression tests with loading in the fiber direction (Schultheisz and Waas 1996). Quek et al. (2004a) found that axial bundles in the loading direction experience buckling failure mechanism during compressive test of carbon 2D triaxial braided composites. For fiber kinking failure criterion, Argon (1972) proposed that the shear stress between fibers caused by fiber initial misalignment angle can lead to fiber kinking. Longitudinal compressive failure stress, X_{LC}, can be calculated from longitudinal shear stress, S_L, and fiber initiation misalignment angle, θ_i, as follows:

$$X_{LC} = \frac{S_L}{\theta_i} \qquad (5.101)$$

Dávila further combines Argon's approach and LaRC02/03 matrix failure criterion to obtain a 2D fiber kinking failure criterion with the LaRC02/03 formulation (Dávila and Camanho 2003). Pinho et al. has extended Dávila's work to obtain the 3D fiber kinking failure criterion (Pinho, Iannucci, and Robinson 2006a,b). As mentioned earlier, all fiber compressive failure criterions are based on the assumption that there exists an initial fiber misalignment angle. For 3D four-directional braided composites, the surface of yarn shows undulation due to the squeezing of braiding yarns as shown in Figure 5.4. Meanwhile, the path of fiber undulation has some periodic characteristics, which have been elaborated in Section 5.2.1. Initial fiber misalignment angle is considered in the FE modeling of RUC (Fang, Liang, Lu et al. 2011). The longitudinal fiber–matrix shear failure is judged by Hashin's failure criterion (Hashin 1980). The maximum stress criterion estimates the fiber kinking failure, and the Mohr–Coulomb criterion is adopted for transverse matrix failure (Puck and Schürmann 2002). The corresponding relations between failure modes and their damage initiation criteria can be written as follows.

The longitudinal (L direction) fiber–matrix shear failure criterion, in which longitudinal shear nonlinearity is taken into account, is expressed as

$$f_L = \begin{cases} \left(\dfrac{\sigma_L}{X_{yt}}\right)^2 + \dfrac{\tau_{LT}^2/(2G_{LT}) + (3/4)A\tau_{LT}^4}{S_L/(2G_{LT}) + (3/4)AS_L^4} \\ \qquad + \dfrac{\tau_{ZL}^2/(2G_{ZL}) + (3/4)A\tau_{ZL}^4}{S_L/(2G_{ZL}) + (3/4)AS_L^4} = 1 \quad (\sigma_L \geq 0) \\[2em] \left(\dfrac{\sigma_L}{X_{yc}}\right)^2 + \dfrac{\tau_{LT}^2/(2G_{LT}) + (3/4)A\tau_{LT}^4}{S_L/(2G_{LT}) + (3/4)AS_L^4} \\ \qquad + \dfrac{\tau_{ZL}^2/(2G_{ZL}) + (3/4)A\tau_{ZL}^4}{S_L/(2G_{ZL}) + (3/4)AS_L^4} = 1 \quad (\sigma_L < 0) \end{cases} \tag{5.102}$$

where σ_L, τ_{LT}, and τ_{ZL} are one longitudinal and two shear stresses, respectively. The fiber kinking criterion is

$$f_K = \left(\frac{\sigma_L}{X_{yc}}\right)^2 = 1 \quad (\sigma_L < 0) \tag{5.103}$$

The transverse matrix failure criterion can be categorized into two forms in terms of the magnitude of normal stress, σ_N. In addition, it is assumed that the transverse tensile strength, Y_T, and longitudinal shear strength, S_L, of the yarn is not influenced by the angle ϕ_0 of fracture plane as shown in Figure 5.8b, i.e.

$$f_T = \begin{cases} \left(\dfrac{\tau_T}{S_T - \mu_T\sigma_N}\right)^2 + \left(\dfrac{\tau_L}{S_L - \mu_L\sigma_N}\right)^2 = 1 & (\sigma_N \leq 0) \\[1.5em] \left(\dfrac{\sigma_N}{Y_{yt}}\right)^2 + \left(\dfrac{\tau_L}{S_L - \mu_L\sigma_N}\right)^2 + \left(\dfrac{\tau_T}{S_T - \mu_T\sigma_N}\right)^2 = 1 & (\sigma_N > 0) \end{cases} \tag{5.104}$$

where σ_N, τ_T, and τ_L are the normal and two shear stresses in the failure plane as shown in Figure 5.8b. It is noted that μ_T and μ_L are the transverse and longitudinal friction coefficients in the fracture plane, respectively.

The transverse friction coefficient can be associated with the angle of fracture plane in pure compression. It is expressed as

$$\tan(2\phi_0) = -\frac{1}{\mu_T} \tag{5.105}$$

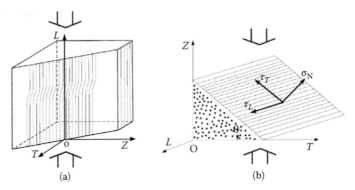

FIGURE 5.8 Different compressive failure modes: (a) longitudinal compressive kinking failure and (b) transverse compressive failure.

The typical values of the angle of fracture plane for unidirectional fiber composites under pure compressive loading, as reported in the literature (González and Lorca 2007), are in the range of 50°–60°. The value of angle ϕ_0 can be assumed to be 53° for simplicity.

Furthermore, S_T can be established from the relation with ϕ_0 and Y_{yc}:

$$S_T = Y_{yc} \cos\phi_0 \left(\sin\phi_0 + \frac{\cos\phi_0}{\tan 2\phi_0} \right) \tag{5.106}$$

The longitudinal friction coefficient μ_L in Equation 5.104 can be obtained from the transverse friction coefficient and the longitudinal and transverse shear strengths, as proposed by Puck and Schürmann, that is,

$$\frac{\mu_L}{S_L} = \frac{\mu_T}{S_T} \tag{5.107}$$

When the braiding yarn is in a complex stress state, the angle of action plane, ϕ, whose normal is rotating in the TZ plane, as shown in Figure 5.8b, will be a variable. The values of σ_N, τ_T, and τ_L all change with ϕ in a stress state. Therefore, Equation 5.104 is a function with respect to ϕ in a stress state, which can be expressed as

$$f_T(\phi) = \begin{cases} \left(\dfrac{\tau_T(\phi)}{S_T - \mu_T\sigma_N(\phi)} \right)^2 + \left(\dfrac{\tau_L(\phi)}{S_L - \mu_L\sigma_N(\phi)} \right)^2 = 1 & (\sigma_N(\phi) \leq 0) \\[4mm] \left(\dfrac{\sigma_N(\phi)}{Y_{yt}} \right)^2 + \left(\dfrac{\tau_L(\phi)}{S_L - \mu_L\sigma_N(\phi)} \right)^2 + \left(\dfrac{\tau_T(\phi)}{S_T - \mu_T\sigma_N(\phi)} \right)^2 = 1 & (\sigma_N(\phi) > 0) \end{cases} \tag{5.108}$$

When ϕ changes in an interval [−90°, 90°], there must exist a maximum value of $f_T(\phi)$. The fracture angle ϕ_0 is determined by searching the maximum value of $f_T(\phi)$:

$$f_T(\phi_0) = \max f_T(\phi)$$
$$\text{if} \quad \max f_T(\phi) \geq 1 \tag{5.109}$$

To find the maximum value of $f_T(\phi)$ in an interval $[-90°, 90°]$, the extended golden section search method, which can avoid unnecessary iterations to obtain an accurate result (Wiegand, Petrinic, and Elliott 2008), can be adopted. First, the golden section search is utilized several times to obtain a subinterval $[\phi_1, \phi_3]$ which contains the maximum value of $f_T(\phi)$. Then, an arbitrary point ϕ_2 between ϕ_1 and ϕ_3 is chosen to generate a parabola formula by interpolation. The maximum value of the parabola formula is close to the maximum value of the function $f_T(\phi)$. Therefore, the fracture angle ϕ_0 can be obtained by

$$\phi_0 \approx \phi_2 - \frac{(\phi_2 - \phi_1)^2(f_T(\phi_2) - f_T(\phi_3)) - (\phi_2 - \phi_3)^2(f_T(\phi_2) - f_T(\phi_1))}{2((\phi_2 - \phi_1)(f_T(\phi_2) - f_T(\phi_3)) - (\phi_2 - \phi_3)(f_T(\phi_2) - f_T(\phi_1)))} \quad (5.110)$$

if $f_T(\phi_0) \geq 1$.

5.4.2.2 Damage Variables and Constitutive Tensor

Three damage indicators can be introduced for the aforementioned damage modes: d_K, d_T, and d_L, corresponding to fiber buckling damage, transverse matrix damage, and longitudinal fiber–matrix shear damage, respectively. The mechanical behavior of braiding yarn is then treated as orthotropic at any damage state. In 2D plane damage models, it is assumed that the elastic modulus, E_i $(i = L, T, Z)$, and shear modulus, G_i $(i = LT, ZL, TZ)$, are influenced by the existence of damage, whereas the variation of v_{LT}/E_L, v_{ZL}/E_Z, and v_{TZ}/E_T can be ignored (Matzenmiller, Lubliner, and Taylor 1995). The degradation level of longitudinal modulus E_L and shear moduli G_{LT} and G_{ZL} is quantified by damage indicators d_K and d_L. E_T, E_Z, G_{LT}, G_{ZL}, and G_{TZ} are degraded by the transverse matrix damage indicator d_T. However, E_T and E_Z are not degraded when the stresses in the T and Z directions of braiding yarn are compressive stresses.

To introduce damage into the constitutive tensor, the damage indicators d_K, d_T, and d_L should be associated with three principal damage indicators. Because the damage indicator d_T is relevant to the fracture angle ϕ_0, the three principal damage indicators of braiding yarn can be constructed as

$$\begin{aligned} \Omega_L &= \max(d_L, d_K) \\ \Omega_T &= d_T \cos\phi_0 \\ \Omega_Z &= d_T \sin\phi_0 \end{aligned} \quad (5.111)$$

To improve the convergence of the numerical algorithm, an artificial Duvaut–Lions viscous model is used (Duvaut and Lions 1972). The time derivatives of the damage variables D_L, D_T, and D_Z can be defined as

$$\dot{D}_I = \frac{1}{\eta_I}(\Omega_I - D_I) \quad I = (L, T, Z) \quad (5.112)$$

where η_I, Ω_I, and D_I are the viscous parameter, unregulated damage variable, and regulated damage variable, respectively, of damage mode I.

Therefore, the constitutive equation including damage for yarn can be expressed as

$$\varepsilon = S(D)\sigma \tag{5.113}$$

where $S(D)$ is the compliance tensor dependence on damage. The unilateral damage is considered in the constitutive equation. The matrix formulation of $S(D)$ is

$$S(D) = \begin{bmatrix} \dfrac{1}{b_L E_L} & 0 & 0 & 0 & 0 & 0 \\[2ex] & \dfrac{1}{b_T E_T} & 0 & 0 & 0 & 0 \\[2ex] & & \dfrac{1}{b_Z E_Z} & 0 & 0 & 0 \\[2ex] & & & \dfrac{1}{b_{LT} G_{LT}} & 0 & 0 \\[2ex] & sym & & & \dfrac{1}{b_{ZL} G_{ZL}} & 0 \\[2ex] & & & & & \dfrac{1}{b_{TZ} G_{TZ}} \end{bmatrix} \tag{5.114}$$

where

$b_L = 1 - D_L$, $b_T = 1 - \langle \sigma_T \rangle D_T$, $b_Z = 1 - \langle \sigma_Z \rangle D_Z$, $b_{LT} = (1 - D_L)(1 - D_T)$, $b_{ZL} = (1 - D_L)(1 - D_Z)$, $b_{TZ} = (1 - D_T)(1 - D_Z)$, and $\langle x \rangle = (x + |x|)/2$.

It can be found that the compliance tensor maintains the symmetrical characteristic, and the different damage modes can be easily distinguished in constructing damage models.

5.4.2.3 Damage Evolution Law

When the damage of braiding yarn is initiated, the development of damage indicator will be controlled by a damage evolution law. The damage evolution law is associated with the element's strains, its material fracture energy, and its characteristic length, which has been established in plane problems (Bažant and Oh 1983; Lapczyk and Hurtado 2007; Maimí et al. 2007) to alleviate the mesh dependence. It is known that the element's dissipated energy equals its elastic energy when the element has been broken. It can be written as

$$\frac{1}{2} \varepsilon_{I,f} \sigma_{I,f} l^3 = G_I l^2, \quad I = L, T, K \tag{5.115}$$

where l is the characteristic length of the element, which is the cube root of element volume in the spatial problem, and $\sigma_{I,f}$, $\varepsilon_{I,f}$, and G_I are the equivalent peak stress, equivalent peak strain, and fracture energy, respectively, of failure mode I. It is assumed for simplicity that the value of fracture energy of one type of fracture mode is constant, whereas the value of failure strain will change with element size.

The equivalent displacement is defined as

$$X_{eq}^I = \varepsilon_I l, \quad I = L, T, K \tag{5.116}$$

where ε_I is the equivalent strain.

Using Equations 5.115 and 5.116, the equivalent displacement at failure point can be written as

$$X_{eq}^{If} = \varepsilon_{I,f} l = 2 G_I / \sigma_{I,f}, \quad I = L, T, K \tag{5.117}$$

The initial equivalent displacement for damage mode I is the value of Equation 5.116 when the initiation damage criterions are equal to one. It can be expressed as

$$X_{eq}^{Ii} = X_{eq}^I \big|_{f_I = 1} \tag{5.118}$$

The equivalent stress and equivalent strain for the aforementioned three damage modes can be defined, as listed in Table 5.5. The fracture toughness for the aforementioned failure modes in Equation 5.115 needs to be determined. Because it is difficult to obtain the fracture toughness of braiding yarn by experiments, it seems reasonable that the fracture toughness of braiding yarn is substituted by the fracture toughness of unidirectional fiber reinforced composites that have the same constituents of braiding yarn of braided composites. In the absence of

---

TABLE 5.5

Equivalent Displacements and Stresses Corresponding to Different Failure Modes

Failure Modes	Equivalence Displacement	Equivalence Stress
L	$X_{eq}^L = l\sqrt{\varepsilon_L^2 + \gamma_{LT}^2 + \gamma_{ZL}^2}$	$l(\sigma_L \varepsilon_L + \sigma_{LT}\varepsilon_{LT} + \sigma_{ZL}\varepsilon_{ZL})/X_{eq}^L$
K	$X_{eq}^K = l\langle -\varepsilon_L \rangle^a$	$l\langle -\sigma_L \rangle\langle -\varepsilon_L \rangle / X_{eq}^K$
T	$X_{eq}^T = l\sqrt{\left(\langle \varepsilon_N \rangle\right)^2 + \gamma_L^2 + \gamma_T^2}$	$l(\langle \sigma_N \rangle\langle \varepsilon_N \rangle + \tau_L \gamma_L + \tau_T \gamma_T)/X_{eq}^T$

[a] $\langle x \rangle = (x + |x|)/2$.

experimental data, the fracture toughness of the failure modes (L, T, and K) for braiding yarn can be obtained from the weight average of fracture toughness for simple loading failure:

$$G_L = \Gamma_{L1} \left(\frac{\langle \sigma_L \rangle}{\sigma_{eq}^L} \right)^2 + \Gamma_{L1-} \left(\frac{\langle -\sigma_L \rangle}{\sigma_{eq}^L} \right)^2 + \Gamma_2 \left(\frac{\tau_{LT}}{\sigma_{eq}^L} \right)^2 + \Gamma_2 \left(\frac{\tau_{TZ}}{\sigma_{eq}^L} \right)^2$$

$$G_K = \Gamma_{L1-} \left(\frac{\langle -\sigma_L \rangle}{\sigma_{eq}^K} \right)^2 \tag{5.119}$$

$$G_T = \Gamma_{T1} \left(\frac{\langle \sigma_N \rangle}{\sigma_{eq}^N} \right)^2 + \Gamma_2 \left(\frac{\tau_T}{\sigma_{eq}^T} \right)^2 + \Gamma_2 \left(\frac{\tau_L}{\sigma_{eq}^T} \right)^2$$

where the fracture toughness values Γ_{L1} and Γ_{L1-} correspond to tensile and compressive loads, respectively, acting alone in the L direction. The fracture toughness Γ_{T1} is associated with the pure tension in the T direction. Γ_2 is mode II intralaminar fracture toughness.

The damage evolvement law of the damage variable $d_I (I = L, T, K)$ for different failure modes can be written as

$$d_I = \frac{X_{eq}^{lf} (X_{eq}^I - X_{eq}^{li})}{X_{eq}^I (X_{eq}^{lf} - X_{eq}^{li})}, \quad (I = L, T, K) \tag{5.120}$$

The damage variable can be substituted in Equation 5.111 to update the development of damage in the damage model.

5.4.3 SIMULATION PROCESS

In the RUC of braided composites, it can be assumed that the yarn–matrix and yarn–yarn interfaces are connected perfectly when the initial interface deficiencies in the RUC are not considered. Meanwhile, the interface damage of yarn–matrix and yarn–yarn can also be taken into account by no-thickness interface element (Fang, Liang, Wang et al. 2011). The aforementioned anisotropic damage model is carried out in material constitutive equation by the user subroutine UMAT of ABAQUS nonlinear finite element codes. To obtain a fast convergence rate of the numerical algorithm for the nonlinear problem, it is necessary to calculate precisely the material tangent constitutive tensor, C_T:

$$\dot{\sigma} = C_T \dot{\varepsilon}, \quad C_T = C + \left(\sum_I \frac{\partial C}{\partial D_I} \frac{\partial D_I}{\partial \Omega_I} \frac{\partial \Omega_I}{\partial d_I} \frac{\partial d_I}{\partial \varepsilon} \right) : \varepsilon, \quad (I = L, T, Z) \tag{5.121}$$

The process of simulation can be shown in the flowchart in Figure 5.9. It has the following stages:

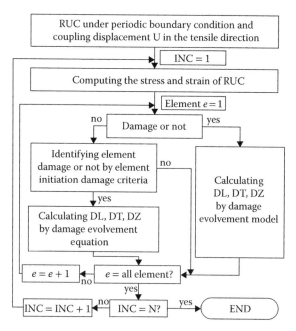

FIGURE 5.9 Flowchart of progressive damage analysis of incremental finite element method.

1. The local stresses and strains of yarns and matrix in the RUC are obtained by Newton–Raphson incremental method after the periodic boundary conditions are applied.
2. The stresses, strains, and damage variables of integral points in every element are introduced to subroutine UMAT. The integral point of an element is identified, that is, whether it has been damaged or not, based on its damage variable values.
3. Damage variables of the element integral point are updated by Equations 5.111, 5.112, and 5.120 if damage appears. Equations 5.102, 5.103, and 5.104 are used to judge whether the element integral point has been damaged, if damage did not appear.
4. The tangent constitutive tensor is updated. The values of stress and strain are calculated.
5. The analysis goes to the next increment after the convergence criterion is reached by several equilibrium iterations.

5.4.4 Damage Development in Braided Composites

The stress–strain curves are different for braided composites with different braid angles. In the damage micromechanical model, the yarn kinking failure mode is considered but does not appear because the longitudinal fiber–matrix shear damage mode appears early in the calculation process. Therefore, there are two damage and failure modes, longitudinal fiber–matrix shear damage and transverse matrix

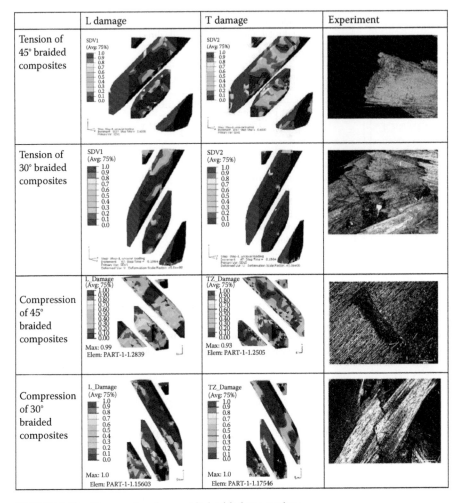

FIGURE 5.10 Damage development in braided composites.

damage, for braiding yarn in the simulation. Figure 5.10 shows the damage states of the inner one-directional braiding yarns within a RUC corresponding to L and T directional damage. The longitudinal fiber–matrix shear damage and transverse matrix damage for 45° braided composites under tensile and compressive loading appear a little earlier in the stress concentration zones where braiding yarns contact each other. The longitudinal damage zones become larger and join neighboring zones, which are perpendicular to the longitudinal direction of braiding yarn. The transverse damage zones become larger and exhibit dispersed distribution. The damage path of longitudinal damage predicted by the model is observed to be the same as the microscopic experimental graphs of braiding yarn in 45° braided composites, as shown in Figure 5.10. In addition, the longitudinal fiber–matrix shear and transverse matrix damage development of one-directional braiding yarns of 30° braided composites under tensile and compressive loading is provided in Figure 5.10.

It can be found that the transverse matrix damage apparently emerges in the zones where yarns contact each other. The longitudinal fiber–matrix shear damage forms a damage strap that has a similar path as experimentally observed and shown in Figure 5.10. Above all, it is clear that this approach has captured the damage and failure mechanisms observed in experiments and has provided a valid method to study the failure process and to predict the compressive strength of braided composites for engineering application.

5.5 STRENGTH AND FAILURE ANALYSIS

The damage and failure analysis of 3D braided composites has been conducted by analytical methods in recent years (Huang 2002; Aggarwal, Ramakrishna, and Ganesh 2002; Tabiei and Ivanov 2004). These methods can efficiently calculate the strength of the textile composites with complex structures. However, detailed geometrical shape and space configurations have been neglected in these models. Finite element analysis is one of the most powerful tools to study the mechanical properties of 3D braided composites (Quek et al. 2004b; Wang et al. 2012) in which the elaborated braid structures in mesoscale can be taken into account. For 3D braided composites, there are complex damage modes in the failure process. And it is difficult to conduct experimental studies for 3D braided composites due to the absence of experimental standards. Therefore, it becomes necessary to predict the strength of 3D braided composites by the numerical models for improving the knowledge of failure process.

5.5.1 LOADING CONDITIONS AND BIAXIAL FAILURE STRESS

The detailed periodic boundary conditions applied on the RUC are provided in Section 5.3.3. The tensile strength (X_T, Y_T, and Z_T), compressive strength (X_C, Y_C, and Z_C), and shear strength (Q, R, and S) can be obtained by simple tensile, compressive, and shear loadings, respectively. The uniaxial and biaxial boundary conditions

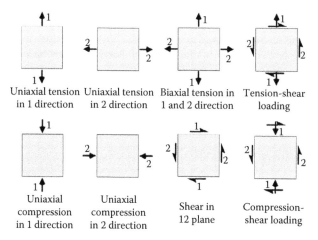

FIGURE 5.11   Different boundary conditions.

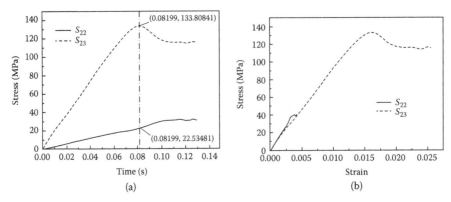

FIGURE 5.12 Stress–time and stress–strain curves of braided composites under tension in 2 directions and shear in 23 plane: (a) stress–time curves and (b) stress–strain curves.

of the braided composites are given in Figure 5.11. The strength values of the braided composites under biaxial loading, which are difficult to be measured by experimental methods, can be obtained by the numerical model of RUC subjected to biaxial loadings. The combined stress can be recognized as a failure point of the braided composite as long as one loading direction stress has reached the peak value. In Figure 5.12, the stress versus time and stress versus strain curves of braided composites under biaxial loading ($\sigma_{22} - \tau_{23}$) are provided. It can be found that the stress S_{22} in the 2 direction is still increasing when the shear stress S_{23} has reached the peak value. Therefore, the point (22.53, 133.8) corresponding to (S_{22}, S_{23}) is a failure point of the braided composites.

5.5.2 Strength Criterion

It is important to establish strength criteria for the application of 3D braided composites in engineering. There exist a number of anisotropic failure criteria for fiber reinforced composites can be adopted, such as limit criteria (maximum stress, maximum strain), interactive criteria (Tsai–Hill, Tsai–Wu), and failure mode–based criteria (Hashin–Rotem, Cuntze). It should be validated whether these failure criteria can be applied for 3D braided composites or not.

Tsai–Wu quadratic polynomial criterion, which is most extensively used in the design of composites (Tsai and Wu 1971), can be expressed as

$$
\begin{aligned}
\sigma_{11}\left(\frac{1}{X_T}+\frac{1}{X_C}\right)+\sigma_{22}\left(\frac{1}{Y_T}+\frac{1}{Y_C}\right)+\sigma_{33}\left(\frac{1}{Z_T}+\frac{1}{Z_C}\right) \\
-\frac{\sigma_{11}^2}{X_T X_C}-\frac{\sigma_{22}^2}{Y_T Y_C}-\frac{\sigma_{33}^2}{Z_T Z_C}+\frac{\tau_{23}^2}{Q^2}+\frac{\tau_{13}^2}{R^2}+\frac{\tau_{12}^2}{S^2} \\
+2F_{12}\sigma_{11}\sigma_{22}+2F_{13}\sigma_{11}\sigma_{33}+2F_{23}\sigma_{22}\sigma_{33}=1
\end{aligned} \tag{5.122}
$$

where compressive strength is a negative value. The interaction coefficients F_{12}, F_{13}, and F_{23} can affect the direction of long axis of the failure ellipse. In view of the border of failure curve, the data range of the interaction coefficient is

$$-\sqrt{F_{ii}F_{jj}} < F_{ij} < \sqrt{F_{ii}F_{jj}} \quad \text{or} \quad -1 < F_{ij}/\sqrt{F_{ii}F_{jj}} < 1 \quad (i,j = 1,2 \text{ or } 1,3 \text{ or } 2,3) \quad (5.123)$$

It is pointed out that the values of the interaction coefficients should be equal to a negative value. Actually, it is difficult to determine the values of the interaction coefficients. Tsai and Hahn have converted the Tsai–Wu polynomial equation into the form of isotropic yielding criterion. Compared with Mises criterion, the interaction coefficients can be written in the following forms:

$$F_{12} = -\frac{1}{2\sqrt{X_T X_C Y_T Y_C}}, \quad F_{13} = -\frac{1}{2\sqrt{X_T X_C Z_T Z_C}}, \quad F_{23} = -\frac{1}{2\sqrt{Y_T Y_C Z_T Z_C}} \quad (5.124)$$

Narayanaswami (Narayanaswami and Adelman 1977) numerically demonstrated that setting the stress interaction coefficient to zero in Tsai–Wu quadratic failure criterion in plane stress analysis results in less than 10% error for all the load cases and materials considered, that is, $F_{ij} = 0$. In fact, the values of the interaction coefficients can be determined theoretically by performing biaxial experiments. For instance, when the stress ratio of biaxial loadings equals the ratio of uniaxial strength values in the 1 and 2 directions, the interaction coefficient F_{12} can be expressed as

$$F_{12} = \frac{1 - (F_{11}(X_T/Y_T)^2 + F_{22})\sigma_{22}^2 - (F_1(X_T/Y_T) + F_2)\sigma_{22}}{2(X_T/Y_T)\sigma_{22}^2} \quad (5.125)$$

where σ_{22} is the failure stress in the 2 direction. $F_1 = \left(\dfrac{1}{X_T} + \dfrac{1}{X_C}\right)$ and $F_2 = \left(\dfrac{1}{Y_T} + \dfrac{1}{Y_C}\right)$.

5.5.3 FAILURE LOCI OF BRAIDED COMPOSITES

In certain biaxial loading conditions, a series of failure points can be obtained by the numerical model with different proportions of loading. Thus, the failure points corresponding to tension–tension and tension–shear loadings that occur in a majority of virtual tests can be calculated. Meanwhile, the 2D Tsai–Wu and Tsai–Hahn failure envelopes can be given based on the failure stresses corresponding to simple loads. The numerical failure points of tension–tension $(\sigma_{22} - \sigma_{33})$ are very close to the failure envelopes of Tsai–Hahn and Tsai–Wu criteria, as shown in Figure 5.13a. On the contrary, the biaxial $(\sigma_{11} - \sigma_{22})$ numerical results are far different from those obtained by these criteria. It can be found, as shown in Figure 5.13b, that the long axis of the hand-drawn ellipse based on the numerical failure points has a certain angle with the horizontal axis. This angle is associated with the interaction coefficient between the two normal stresses. At the same time, the uniaxial failure point stress in the 2 direction is nearly equal to that in the 3 direction, whereas the uniaxial failure point stress in the 1 direction is much larger than those in the 2 and 3 directions.

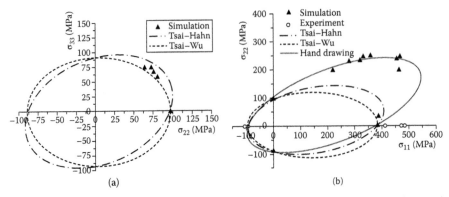

FIGURE 5.13 Tension–tension failure loci: (a) tension–tension in 2 and 3 directions and (b) tension–tension in 1 and 2 directions.

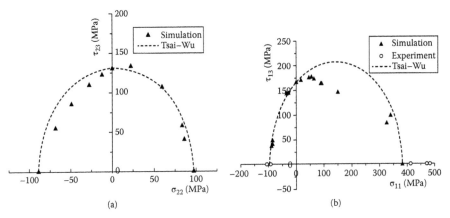

FIGURE 5.14 Tension–shear and compression–shear failure loci: (a) 2 and 23 directions and (b) 1 and 13 directions.

Therefore, the interaction coefficients between two normal failure stresses should be considered seriously when the difference between two normal failure stress values is large. Figure 5.14 shows the tension–shear and compression–shear ($\sigma_2 - \tau_{23}$ and $\sigma_1 - \tau_{13}$) failure loci. The failure points of $\sigma_2 - \tau_{23}$ are very close to the failure locus obtained by Tsai–Wu criterion, whereas the numerical results of tension–shear $\sigma_1 - \tau_{13}$ are far different from the results calculated by Tsai–Wu criterion. Therefore, it is required to develop a failure criterion for 3D braided composites.

It can be found from the aforementioned numerical results that these numerical failure points exhibit some rules. So, a large number of failure points of the braided composites under simple and complex loadings can be obtained finally by numerical calculation. Then, the failure loci of the braided composites can be fitted with these failure points. Further, it is helpful to establish a failure criterion for the braided composites.

5.6 SUMMARY

The mechanical properties of braided composites can be obtained directly by FEM combined with RUC. However, the periodic boundary conditions imposed on RUC should be established correctly. Using this approach, the effective properties of braided composites can be calculated accurately. These methods can also be used for the computation of effective properties of other textile composites. In the damage and failure analysis of braided composites, the failure stresses and damage mechanisms of constituents (braiding yarn and matrix) can be considered to describe the failure process of the braided composites. A large number of failure points for the braided composites corresponding to simple and biaxial loadings can be calculated. The failure loci of the braided composites are provided by numerical failure points. The failure criterion of braided composites can be established and developed by utilizing progressive damage analysis models.

There are some points which can be developed further. First, if RUC is used to evaluate the mechanical properties of the braided composites, the braided composites should satisfy the periodic characteristics. This assumption may be suitable for a thick and large component manufactured from braided composites. But for a thin-walled component with a complex shape, the periodic characteristic may not be satisfactory and the mechanical properties of the component cannot be obtained by RUC. Therefore, a model is required to predict realistic textile geometry when considering textile composites with complex internal architectures. Second, the matrix configuration within the textile composites is complex. Thus, the in situ mechanical properties of constituents within textile composites are difficult to determine. It may be an effective way to obtain the mechanical properties of constituents through reverse calculation. Third, the initial damage criterion is mainly chosen manually to influence the results of damage and failure analysis. Therefore, the initial damage criterion should be determined analytically and physically. Fourth, the theoretical and numerical model for other mechanical properties of textile composites, such as fatigue, impact, and oxidation ablation, which are mainly studied by experiment, should be developed. Last but not least, the macroscopic phenomenological failure theory for certain textile composites, which may include the characteristic parameter to describe the mesostructure, should be established to satisfy the requirements from engineering applications.

REFERENCES

Aboudi, J. 1991. *Mechanics of Composite Materials: A Unified Micromechanical Approach.* New York: Elsevier.

Aggarwal, A., S. Ramakrishna, and V. K. Ganesh. 2002. "Predicting the strength of diamond braided composites." *Journal of Composite Materials* 36 (5):625–643. doi: 10.1177/0021998302036005487.

Argon, AS. 1972. "Fracture of composites." In *Treatise on Materials Science and Technology*, edited by H. Herman, 79–114. New York: Academic Press.

Babuška, I. 1976. "Homogenization approach in engineering." In *Computing Methods in Applied Sciences and Engineering*, 137–153. Berlin, Germany: Springer.

Bažant, Z. P., and B. H. Oh. 1983. "Crack band theory for fracture of concrete." *Matériaux et Construction* 16 (3):155–177. doi: 10.1007/BF02486267.

Belytschko, T., and T. Black. 1999. "Elastic crack growth in finite elements with minimal remeshing." *International Journal for Numerical Methods in Engineering* 45 (5):601–620.

Belytschko, T., N. Moës, S. Usui, and C. Parimi. 2001. "Arbitrary discontinuities in finite elements." *International Journal for Numerical Methods in Engineering* 50 (4):993–1013.

Belytschko, T., C. Parimi, N. Moës, N. Sukumar, and S. Usui. 2003. "Structured extended finite element methods for solids defined by implicit surfaces." *International Journal for Numerical Methods in Engineering* 56 (4):609–635.

Budiansky, B., and N. A. Fleck. 1993. "Compressive failure of fibre composites." *Journal of the Mechanics and Physics of Solids* 41 (1):183–211.

Byun, Joon-Hyung, and Tsu-Wei Chou. 1996. "Process-microstructure relationships of 2-step and 4-step braided composites." *Composites Science and Technology* 56 (3):235–251.

Chen, L., X. M. Tao, and C. L. Choy. 1999a. "On the microstructure of three-dimensional braided preforms." *Composites Science and Technology* 59 (3):391–404.

Chen, L., X. M. Tao, and C. L. Choy. 1999b. "Mechanical analysis of 3-D braided composites by the finite multiphase element method." *Composites Science and Technology* 59 (16):2383–2391.

Daniels, H. E. 1945. "The statistical theory of the strength of bundles of threads. I." *Proceedings of the Royal Society of London. Series A. Mathematical and Physical Sciences* 183 (995):405–435.

Dávila, C. G., and P. P. Camanho. 2003. "Failure criteria for FRP laminates in plane stress." NASA/TM-2003-212663.

Dolbow, J., and T. Belytschko. 1999. "A finite element method for crack growth without remeshing." *International Journal for Numerical Methods in Engineering* 46 (1):131–150.

Duvaut, G., and J. L. Lions. 1972. *Les Inéquations en Mécanique et en Physique*. Paris, France: Dunod.

Fang, G. D., and J. Liang. 2011. "A review of numerical modeling of three-dimensional braided textile composites." *Journal of Composite Materials* 45 (23):2415–2436. doi: 10.1177/0021998311401093.

Fang, G. D., J. Liang, Q. Lu, B. L. Wang, and Y. Wang. 2011. "Investigation on the compressive properties of the three dimensional four-directional braided composites." *Composite Structures* 93 (2):392–405. doi: 10.1016/j.compstruct.2010.09.002.

Fang, G. D., J. Liang, and B. L. Wang. 2009. "Progressive damage and nonlinear analysis of 3D four-directional braided composites under unidirectional tension." *Composite Structures* 89 (1):126–133. doi: 10.1016/j.compstruct.2008.07.016.

Fang, G. D., J. Liang, B. L. Wang, and Y. Wang. 2011. "Effect of interface properties on mechanical behavior of 3D four-directional braided composites with large braid angle subjected to uniaxial tension." *Applied Composite Materials* 18 (5):449–465. doi: 10.1007/s10443-010-9175-6.

Fang, G. D., J. Liang, Y. Wang, and B. L. Wang. 2009. "The effect of yarn distortion on the mechanical properties of 3D four-directional braided composites." *Composites Part A: Applied Science and Manufacturing* 40 (4):343–350. doi: 10.1016/j.compositesa.2008.12.007.

González, C., and J. L. Lorca. 2007. "Mechanical behavior of unidirectional fiber-reinforced polymers under transverse compression: microscopic mechanisms and modeling." *Composites Science and Technology* 67 (13):2795–2806.

Green, S. D., A. C. Long, B. S. F. El Said, and S. R. Hallett. 2014. "Numerical modelling of 3D woven preform deformations." *Composite Structures* 108:747–756.

Halpin, J. C. 1992. *Primer on Composite Materials Analysis* (revised). Lancaster, PA: CRC Press.

Hashin, Zvi. 1980. "Failure criteria for unidirectional fiber composites." *Journal of Applied Mechanics* 47 (2):329–334.

Hearle, J. W. S., P. Grosberg, and S. Backer. 1969. *Structural Mechanics of Fibers, Yarns, and Fabrics*. New York: John Wiley & Sons, Inc.

Huang, Zheng-Ming. 2002. "On a general constitutive description for the inelastic and failure behavior of fibrous laminates—part II: laminate theory and applications." *Computers & Structures* 80 (13):1177–1199. doi: http://dx.doi.org/10.1016/S0045-7949(02)00075-5.

Iarve, E. V, D. H. Mollenhauer, E. G. Zhou, T. Breitzman, and T. J. Whitney. 2009. "Independent mesh method-based prediction of local and volume average fields in textile composites." *Composites Part A: Applied Science and Manufacturing* 40 (12):1880–1890.

Ishikawa, T., and T-W Chou. 1982. "Stiffness and strength behaviour of woven fabric composites." *Journal of Materials Science* 17 (11):3211–3220.

Jiang, Wen-Guang, S. R. Hallett, and M. R. Wisnom. 2008. "Development of domain superposition technique for the modelling of woven fabric composites." In *Mechanical Response of Composites*, edited by Camanho PP, Dávila CG, Pinho ST, and Remmers JC. 281–291. Berlin, Germany: Springer-Verlag.

Kim, H. J., and C. C. Swan. 2003. "Voxel-based meshing and unit-cell analysis of textile composites." *International Journal for Numerical Methods in Engineering* 56 (7):977–1006.

Lapczyk, I., and J. A. Hurtado. 2007. "Progressive damage modeling in fiber-reinforced materials." *Composites Part A: Applied Science and Manufacturing* 38 (11):2333–2341. doi: http://dx.doi.org/10.1016/j.compositesa.2007.01.017.

Li, S. 2008. "Boundary conditions for unit cells from periodic microstructures and their implications." *Composites Science and Technology* 68 (9):1962–1974.

Li, S., and A. Wongsto. 2004. "Unit cells for micromechanical analyses of particle-reinforced composites." *Mechanics of Materials* 36 (7):543–572.

Liang, J., S. Du, and J. Han. 1997. "Effective elastic properties of three dimensional braided composites with matrix microcracks." *Fuhe Cailiao Xuebao/Acta Materiae Compositae Sinica* 14 (1):101–107.

Liu, Z., Z. Lu, and Z. Yang. 2006. "Geometrical characteristics of structural model for 3-D braided composites." *Journal of Beijing University of Aeronautics and Astronautics* 1:021.

Lomov, S. V., R. S. Parnas, S. B. Ghosh, I. Verpoest, and A. Nakai. 2002. "Experimental and theoretical characterization of the geometry of two-dimensional braided fabrics." *Textile Research Journal* 72 (8):706–712.

Lomov, S. V., and I. Verpoest. 2003. "WiseTex-virtual textile reinforcement software." 48th International SAMPE symposium proceedings, Long Beach, USA.

Ma, Chang-Long, Jenn-Ming Yang, and Tsu-Wei Chou. 1986. "Elastic stiffness of three-dimensional braided textile structural composites." Composite Materials: Testing and Design (Seventh Conference). ASTM STP 893, Whitney JM, Ed., American Society for testing and Materials, Philadelphia, pp. 404–421.

Maimí, P., P. P. Camanho, J. A. Mayugo, and C. G. Dávila. 2007. "A continuum damage model for composite laminates: part II–computational implementation and validation." *Mechanics of Materials* 39 (10):909–919. doi: http://dx.doi.org/10.1016/j.mechmat.2007.03.006.

Matzenmiller, A. L. J. T. R., J. Lubliner, and R. L. Taylor. 1995. "A constitutive model for anisotropic damage in fiber-composites." *Mechanics of Materials* 20 (2):125–152.

Miao, Y., E. Zhou, Y. Wang, and B. A. Cheeseman. 2008. "Mechanics of textile composites: micro-geometry." *Composites Science and Technology* 68 (7):1671–1678.

Morais, Alfredo Balacó de. 2000. "Transverse moduli of continuous-fibre-reinforced polymers." *Composites Science and Technology* 60 (7):997–1002.

Narayanaswami, R., and H. M. Adelman. 1977. "Evaluation of the tensor polynomial and Hoffman strength theories for composite materials." *Journal of Composite Materials* 11 (4):366–377. doi: 10.1177/002199837701100401.

Pastore, C. M. 2000. "Opportunities and challenges for textile reinforced composites." *Mechanics of Composite Materials* 36 (2):97–116.

Pastore, C. M., and F. K. Ko. 1990. "Modelling of textile structural composites part i: processing-science model for three-dimensional braiding." *Journal of the Textile Institute* 81 (4):480–490.

Pinho, S. T., L. Iannucci, and P. Robinson. 2006a. "Physically-based failure models and criteria for laminated fibre-reinforced composites with emphasis on fibre kinking: part I: development." *Composites Part A: Applied Science and Manufacturing* 37 (1):63–73.

Pinho, S. T., L. Iannucci, and P. Robinson. 2006b. "Physically based failure models and criteria for laminated fibre-reinforced composites with emphasis on fibre kinking. part II: FE implementation." *Composites Part A: Applied Science and Manufacturing* 37 (5):766–777.

Puck, A., and H. Schürmann. 2002. "Failure analysis of FRP laminates by means of physically based phenomenological models." *Composites Science and Technology* 62 (12):1633–1662.

Quek, S. C., A. M. Waas, K. W. Shahwan, and V. Agaram. 2004a. "Compressive response and failure of braided textile composites: part 1—experiments." *International Journal of Non-Linear Mechanics* 39 (4):635–648.

Quek, S. C., A. M. Waas, K. W. Shahwan, and V. Agaram. 2004b. "Compressive response and failure of braided textile composites: part 2—computations." *International Journal of Non-Linear Mechanics* 39 (4):649–663. doi: http://dx.doi.org/10.1016/S0020-7462(03)00019-2.

Schultheisz, C. R., and A. M. Waas. 1996. "Compressive failure of composites, part I: testing and micromechanical theories." *Progress in Aerospace Sciences* 32 (1):1–42.

Sherburn, M. 2007. "Geometric and mechanical modelling of textiles." PhD Thesis, University of Nottingham, United Kingdom.

Stock, S. R. 1999. "X-ray microtomography of materials." *International Materials Reviews* 44 (4):141–164.

Straumit, I., S. V. Lomov, I. Verpoest, and M. Wevers. 2013. "Determination of local fibers orientation in composite material from micro-CT data." *Proceedings of the 11th International Conference on Textile Composites (TexComp-11)* Leuven, Belgium, 16–20 September 2013.

Sun, X. 2004. "Micro-geometry of 3-D braided tubular preform." *Journal of Composite Materials* 38 (9):791–798.

Tabiei, A., and I. Ivanov. 2004. "Materially and geometrically non-linear woven composite micro-mechanical model with failure for finite element simulations." *International Journal of Non-Linear Mechanics* 39 (2):175–188. doi: http://dx.doi.org/10.1016/S0020-7462(02)00067-7.

Tsai, S. W., and E. M. Wu. 1971. "A general theory of strength for anisotropic materials." *Journal of Composite Materials* 5 (1):58–80. doi: 10.1177/002199837100500106.

Verpoest, I., and S. V. Lomov. 2005. "Virtual textile composites software WiseTex: Integration with micro-mechanical, permeability and structural analysis." *Composites Science and Technology* 65 (15):2563–2574.

Wang, B., G. Fang, J. Liang, and Z. Wang. 2012. "Failure locus of 3D four-directional braided composites under biaxial loading." *Applied Composite Materials* 19 (3–4):529–544. doi: 10.1007/s10443-011-9206-y.

Wang, X. 1999. *Mechanical Analysis and Design of Composite Materials*. Changsha, China: National University of Defence Technology Press.

Wang, Y., and X. Sun. 2001. "Digital-element simulation of textile processes." *Composites Science and Technology* 61 (2):311–319.

Wang, Y. Q., and A. S. D. Wang. 1995a. "Geometric mapping of yarn structures due to shape change in 3-D braided composites." *Composites Science and Technology* 54 (4):359–370.

Wang, Y. Q., and A. S. D. Wang. 1995b. "Microstructure/property relationships in three-dimensionally braided fiber composites." *Composites Science and Technology* 53 (2):213–222.

Whitcomb, J. D., C. D. Chapman, and X. Tang. 2000. "Derivation of boundary conditions for micromechanics analyses of plain and satin weave composites." *Journal of Composite Materials* 34 (9):724–747.

Whitcomb, J., K. Srirengan, and C. Chapman. 1995. "Evaluation of homogenization for global/ local stress analysis of textile composites." *Composite Structures* 31 (2):137–149.

Wiegand, J., N. Petrinic, and B. Elliott. 2008. "An algorithm for determination of the fracture angle for the three-dimensional Puck matrix failure criterion for UD composites." *Composites Science and Technology* 68 (12):2511–2517. doi: http://dx.doi.org/10.1016/j.compscitech.2008.05.004.

Wu, D. L. 1996. "Three-cell model and 5D braided structural composites." *Composites Science and Technology* 56 (3):225–233.

Xia, Z., Y. Zhang, and F. Ellyin. 2003. "A unified periodical boundary conditions for representative volume elements of composites and applications." *International Journal of Solids and Structures* 40 (8):1907–1921.

Xu, J., B. N. Cox, M. A. McGlockton, and W. C. Carter. 1995. "A binary model of textile composites—II. The elastic regime." *Acta Metallurgica et Materialia* 43 (9):3511–3524.

Xu, W., and A. M. Waas. 2014. "A novel shell element for quasi-static and natural frequency analysis of textile composite structures." *Journal of Applied Mechanics* 81 (8):081002. doi: 10.1115/1.4027439.

Yang, Jenn-Ming, Chang-Long Ma, and Tsu-Wei Chou. 1986. "Fiber inclination model of three-dimensional textile structural composites." *Journal of Composite Materials* 20 (5):472–484.

Yuan, Z., and J. Fish. 2008. "Toward realization of computational homogenization in practice." *International Journal for Numerical Methods in Engineering* 73 (3):361–380. doi: 10.1002/nme.2074.

Zeng, T., Dai-ning Fang, L. Ma, and Li-cheng Guo. 2004. "Predicting the nonlinear response and failure of 3D braided composites." *Materials Letters* 58 (26):3237–3241.

Zeng, T., Lin-zhi Wu, and Li-cheng Guo. 2004. "A finite element model for failure analysis of 3D braided composites." *Materials Science and Engineering: A* 366 (1):144–151.

Zhou, G., X. Sun, and Y. Wang. 2004. "Multi-chain digital element analysis in textile mechanics." *Composites Science and Technology* 64 (2):239–244.

Zhu, Y., B. Zhou, G. He, and Z. Zheng. 1989. "A statistical theory of composite materials strength." *Journal of Composite Materials* 23 (3):280–287.

Zohdi, T., M. Feucht, D. Gross, and P. Wriggers. 1998. "A description of macroscopic damage through microstructural relaxation." *International Journal for Numerical Methods in Engineering* 43 (3):493–506.

Zuo, Wei-wei, Lai-yuan Xiao, and Dao-xun Liao. 2007. "Statistical strength analyses of the 3-D braided composites." *Composites Science and Technology* 67 (10):2095–2102.

Section II

Application of Braided
Structures and Composites

6 Applications of Braided Structures in Medical Fields

Wen Zhong

CONTENTS

ABSTRACT

Braided structures are known for their exceptional mechanical properties in their longitudinal directions, and have found various applications in medical implants that require good biostability. The most important braided medical implants available commercially and clinically are sutures and artificial ligaments. Various materials, including biodegradable, nondegradable, natural, and synthetic materials have been used for these implants. This chapter discusses about the materials and structures for sutures and artificial ligaments as well as their performance. Examples of product developments are also provided.

6.1 INTRODUCTION

The "braiding" technique is used to bring about a textile structure by diagonal intersection of yarns or filaments. Two-dimensional (2D) braiding produces flat strip or tape, whereas three-dimensional (3D) braiding provides tubular- or rod-shaped textile structures; 3D braids have been used in the reinforcement structure for composite materials to be applied in such high-end products as those employed in military or aerospace sectors. Braided structures are also found in health-care and medical textile products, such as sutures or ligament prosthesis.

Braided structures are known for their exceptional mechanical properties in their longitudinal directions, including excellent tensile strength and flexibility. As a result, if applied in medical applications including implants, they provide good biostability, which is a term to describe the abilities of a biomaterial to maintain its original dimensions and mechanical and chemical properties during an extended period, or for the rest of a patient's life, under a hostile biological environment [1]. For a bioresorbable implant, such as a resorbable suture or a scaffold for tissue repair, it is important that the process of it being absorbed by the body can be predicted or controlled to optimize the remodeling of host tissues. For a nonabsorbable implant, such as a nonabsorbable suture or ligament prosthesis, it is essential that the implant remains stable and functional during its service life. Since an implant functions under a mechanical load and in an environment of liquids that are chemical in nature, a discussion of its biostability has to be from a physical and chemical standpoint. Biostability of a material/structure may be expressed in different terms for specific end uses. For example, when applied in surgical sutures, the biostability of a suture material may be referred to as tensile strength, elasticity/plasticity, and degradation rate; if used for ligament prosthesis, a material/structure may be evaluated in terms of tensile strength, yield strength, and abrasion resistance.

On the other hand, these braided structures are usually expected to be made from a biocompatible material if used in medical fields, especially if they are for internal uses. In other words, it is desired that such a material elicits benign responses from the cells, tissues, and organs of the patient who has received an implanted foreign material [2]. Biocompatibility involves the interactions between the implanted materials (or a device made from them) and host tissues. All aspects of the material properties, including mechanical, chemical, pharmacological, and surface properties, may impact on such interactions [3,4]. There have been two basic methods of evaluation for the biocompatibility of a materials/structure: in vitro and in vivo. The latter approach refers to procedures in which a material/structure is implanted and tested inside animals or humans. On the other hand, the in vitro (or ex vivo) approach depends on procedures in which the material/structure is cultured outside the human body (usually with viable cells) for examination [5].

Biocompatibility and bioresorbability (or biostability) are the most important criteria for evaluating a biotextile material or any fiber/textile product that will react with a biological system in its end uses, and therefore will be used as the major criteria to evaluate the materials mentioned in this chapter for various applications of braided structures.

6.2 TYPE OF APPLICATIONS

6.2.1 SUTURES

Most frequently used, especially in the process of surgery, sutures help close the wound, a function that will last relatively long until the natural healing process is restored to provide a sufficient level of wound strength. The use of sutures may be dated back to thousands of years ago. Natural materials, including flax, hair, cotton, silk and catgut, had been used for this purpose in the many centuries before the invention and commercialization of synthetic sutures in the 1940s. Suture materials can be evaluated according to the following [6]:

1. Physical properties (biostability).
    a. Physical configuration—whether the material is of a single-strand/monofilament or multistrand/multifilament structure.
    b. Diameter (or caliber).
    c. Capillarity—the capacity of a suture to adsorb or imbibe fluid.
    d. Fluid absorption ability—its capacity to absorb fluid into the fiber material.
    e. Tensile strength (or breaking tenacity, is the maximum stress a material may sustain before breaking).
    f. Knot strength—the force necessary to cause a knot to slip.
    g. Elasticity—its ability to regain its original form and length after being stretched.
    h. Plasticity—its ability to retain its new deformed length and form after being stretched.
    i. Memory—its capacity to return to its original shape upon deformation.

    Appropriate physical properties of sutures are related to their performance. Capillary and fluid absorption may influence a suture's tendency to take up and retain water-borne bacteria, which can induce infection. A multifilament suture can take up more fluid and bacteria than a monofilament suture made of the same polymer material. To maintain its strength during the wound healing process, or, to avoid the risk of early degradation of the suture before the wound heals, a bioabsorbable suture should have predictable degradation and absorption (usually accompanied by a reduction of mechanical strength). A suture should also be flexible to allow easy bending and stretching to accommodate wound edema (accumulation of an excessive amount of watery fluid in cells, tissues, or body cavities) and to recoil to its original length after the wound contracts. Memory can be related to elasticity and plasticity of a suture. A suture material (e.g., nylon) with high memory tends to untie from a knot to result in decreased knot security, whereas a suture material (e.g., silk) with low memory usually has a high knot security.
2. "Handling properties" dictate a suture's pliability (flexibility), knottability, and surface/frictional features. Pliability is the measure of how easily the suture can be handled by a surgeon in a surgical operation, especially how flexible it is to be bent. Braided sutures are usually more pliable than monofilament sutures. Knottability is important because a suture should be tied easily to form a secure and minimized knot. Surface/frictional properties (e.g., coefficient of friction) indicate how easily the suture will slip through the tissue and be knotted.

3. "Tissue response or tissue reaction (biocompatibility)" is, on the other hand, undesirable because they impede wound healing and may cause infection and discomfort. Generally, higher tissue response is associated with lower biocompatibility of the suture material, and a larger quantity of foreign material implanted. Therefore, a desirable suture should demonstrate low tissue response. To minimize tissue response, a suture should have high biocompatibility and the least possible amount of material. Tissue responses to absorbable suture are usually more obvious than those to nonabsorbable sutures, and such responses will persist until the suture is absorbed.

It should be noted that the above-mentioned properties of sutures are not independent but are usually correlated. For example, when other technical parameters are kept identical, the diameter of a suture becomes larger when the tensile strength is greater; but when the flexibility is lower, the tissue response becomes higher. On the other hand, the physical and chemical properties of a biomaterial may lead to a prolonged tissue response. For example, tissue responses can be induced by debris detached from an implant (such as sutures) due to abrasion/wear and by degradation products of a biodegradable biomaterial (e.g., acids produced by the degradation of PGA).

Regarding the structure of the sutures, both monofilament sutures and braided multifilament sutures are commercially and clinically available. Although braided sutures may not have particular advantages over monofilament sutures in terms of wound healing properties, they have outstanding handling properties and flexibility that outweigh the beneficial healing properties of monofilament sutures [7].

Sutures can be grouped into two categories according to their source of raw materials, that is, natural sutures and synthetic sutures. Sutures also vary in their capacity to retain a long-term physical integrity in the body, and can therefore be differentiated into absorbable and nonabsorbable sutures. The various types of sutures are summarized in Table 6.1.

The following are types of sutures that are better known and more often used:

6.2.1.1   Absorbable Sutures

"Catgut," known also as surgical gut, is derived from the small intestine of sheep or cattle. The major component of catgut is strands of highly purified collagen, a fibrous protein. Collagen exists naturally in the form of fiber; in structure, one collagen fibril is composed of three polymer chains in a triple helix, and thousands of overlapped

TABLE 6.1
Types of Sutures

	Natural Sutures	Synthetic Sutures
Absorbable	Catgut	Polyglycolic acid (PGA), polyglactin, polydioxanone (PDO)
Nonabsorbable	Silk	Polyamide (nylon), polyester (Dacron®), polypropylene (PP), polytetrafluoroethylene (PTFE)

fibrils form one collagen fiber (Figure 6.1). Catgut as a traditional suture material has now been largely replaced by synthetic sutures because of several disadvantages. Collagen is a biocompatible and bioresorbable natural polymer abundantly found in the connective tissue of animals. A catgut, however, may contain such contaminants as muscle fibers or mucoproteins, which may cause tissue reaction problems. Catgut can be treated with chromium salt, in a procedure similar to the tanning of leather. The resulted chromic gut is stronger and more resistant to degradation after implantation. Catgut sutures generally are low in mechanical strength. Plain gut starts to degrade in 12 hours after implantation, and it will retain significant strength for only 5–7 days, whereas chromic gut may last twice as long [8]. Both plain and chromic gut sutures completely lose their tensile strength in 2–3 weeks. The complete absorption of a gut suture is unpredictable and may last about 12 weeks. Catgut degrades via degradation or digestion of proteins by enzymes (proteolysis) [6,9]. These sutures also have poor handle and knot security: the knots may harden and traumatize adjacent tissues in the presence of body fluid [10]. Catgut is known to cause high degrees of tissue reaction (the highest among currently available sutures) and impede wound healing.

"Polyglycolic acid" (PGA) (Figure 6.2) is a biodegradable and bioresorbable polyester widely used in biomedical applications. It was first introduced as a material for bioresorbable sutures in the 1970s. PGA has good mechanical strength and a predictable bioresorbability, which is a desirable property for implants. PGA sutures usually contain braided filaments produced via solvent spinning and heat stretching. A highly crystalline material, a PGA suture, has better mechanical properties than a catgut suture. It retains 60% of its original breaking strength at day 7 [11] and 20% of its original strength by day 28 [12]. It is completely dissolved within 90–120 days

FIGURE 6.1 Triple helix structure of collagen.

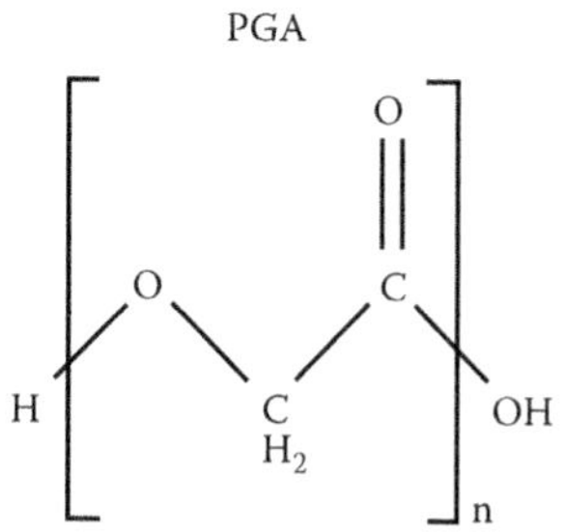

PGA

$$\left[\begin{array}{c} H \end{array} \underset{\substack{C \\ H_2}}{\overset{O}{\underset{\diagdown}{\diagup}}} \overset{\displaystyle O}{\underset{C}{\parallel}} OH \right]_n$$

FIGURE 6.2 Chemical structure of polyglycolic acid.

[13]. PGA is broken down in the body via hydrolysis (i.e., having its long polymer chain disintegrated as a result of the attack of water) in the presence of body fluid. However, there have been concerns that the rapid degradation of PGA via hydrolysis may dramatically decrease the local pH value in the tissue and cause tissue responses if the local area does not have a high buffering capacity or effective mechanisms to remove the metabolic waste [14].

The original PGA suture (i.e., Dexon® S) has a rough surface, making it difficult to handle and secure a knot. The rough surface also increases drag force to the tissue when the suture is pulled through. The next generation of PGA suture (i.e., Dexon Plus) adopts a surface coating to lubricate the suture surface, so that the suture can slip through the tissue more smoothly and allow easier handling and knotting. The coating is soluble in aqueous environment and is eliminated within a few days, resulting in an uncoated suture with the rough surface to retain good knot security [15].

Absorbable PGA sutures are useful for procedures inside the body as no removal procedure is required. However, they are not often used for cutaneous surgery in which the sutures can later be removed because the braided multifilament sutures may allow the penetration of bacteria into the wound by capillary action. Besides, the degradation and absorption rate of the suture is unpredictable if it is not totally buried and in contact with body fluid [6].

"Polyglactin 910" suture (i.e., VICRYL®) was introduced in 1974 as another synthetic and bioresorbable suture. It contains a copolymer of 90% glycolide and 10% lactide, known as poly(lactic-co-glycolic acid), or PLGA, as shown in Figure 6.3. The structure and properties of polyglactin 910 suture are similar to those of the PGA suture. VICRYL is also available in the form of braided filaments produced from solvent spinning. VICRYL retains 8% of its strength by day 28. Complete absorption of VICRYL takes less time (60–90 days) than that of Dexon, because the polylactide blocks render the submicroscopic polymer chains apart and, therefore, cause the copolymer more vulnerable to hydrolytic degradation [13]. Coated VICRYL was also developed to improve the suture's passage through tissue and its knotting properties. Tissue response, handling qualities, and physical properties of coated VICRYL are very similar to those of Dexon Plus. Also, VICRYL suture is more suitable for procedures inside the body than cutaneous surgery.

"Polydioxanone (PDO)" suture (e.g., PDS®) is composed of a polymer synthesized from the monomer, paradioxanone, as shown in Figure 6.4. It is manufactured as a monofilament suture, and therefore has a lesser affinity for bacteria. PDS retains 58% of its tensile strength by day 28 after implantation and is completely

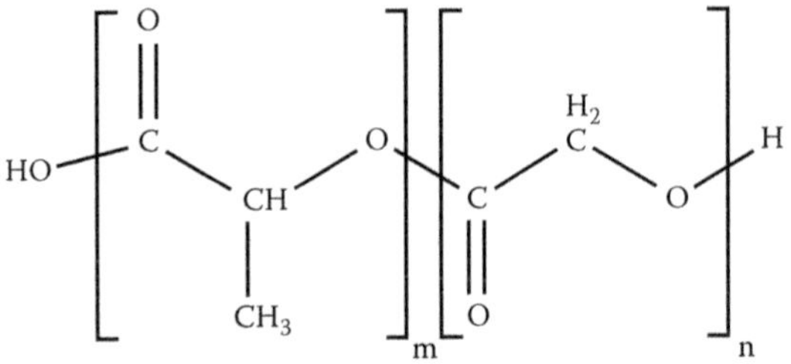

FIGURE 6.3 Chemical structure of PLGA (for polyglactin 910, m:n = 10:90).

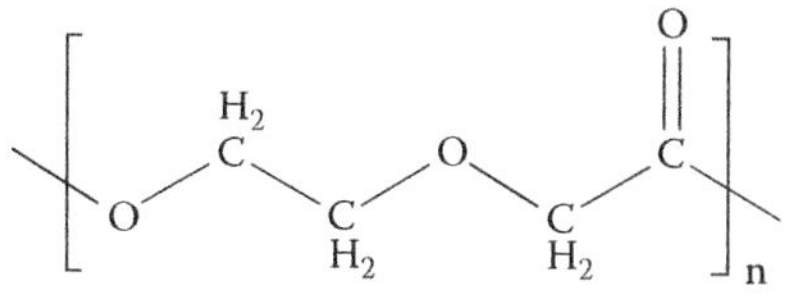

FIGURE 6.4 Chemical structure of polydioxanone.

absorbed in 180 days [16]. Unlike other synthetic absorbable sutures, PDS has a slow degradation profile because of its slow hydrolysis, enabling it to be used for situations where a longer period is required for the wounded skin to regain considerable strength. The PDS suture, as well as its degradation products, imposes minimal tissue response, although the monofilament suture is stiffer than multifilament sutures, thus negatively affecting its handling and knotting properties [9].

6.2.1.2 Nonabsorbable Sutures

"Silk" is a natural protein (fibroin) fiber. Silk is produced by the larva of cultivated or wild silkworms for the construction of their cocoons. Silk is the only natural filament or continuous fiber, with a length up to 600 m and a fine diameter between 12 and 30 μm [17]. The silk fiber is composed of a protein polymer called fibroin, which consists of amino acids arranged in a pleated β-sheet structure (Figure 6.5).

Silk sutures are categorized as a nonabsorbable suture because it takes almost 2 years to be completely absorbed. They are available either in braided or monofilament form. A braided suture is soft and has good handling and knotting properties. However, silk suture is hydrophilic and swells in the presence of body fluid, and the interstices between filaments in braided suture may be covered or penetrated by ingrowth tissues, therefore causing pain during suture removal. A monofilament silk suture, on the other hand, does not allow tissue penetration and ingrowth, thus imposing less pain during removal [6]. The silk suture is weaker than other synthetic sutures, and it is generally believed that it induces a higher level of tissue response than most other sutures except catgut [18]. However, a recent report suggests that the higher level of tissue response may be caused by the residual sericin (glue-like proteins left behind after the degumming and scouring process in silk) on the silk fiber surface, whereas the core silk fibroin fibers exhibit comparable biocompatibility with other commonly used biomaterials [19].

"Polyamide" (nylon) was the first synthetic suture, introduced in 1940. According to the number(s) of carbon atoms in the monomer(s) that form the nylon polymer, the variety of nylon fibers are named as nylon 3, nylon 4, nylon 5, nylon 6, nylon 6,6 (i.e., two six-carbon atom monomers are involved in the polymerization), etc. Among them, nylon 6 and nylon 6,6 (Figure 6.6) have the most frequent textile end uses, including their applications in sutures. Nylon is slowly absorbable, classified as a nonabsorbable suture. It loses its strength by hydrolysis at a rate of 15%–20% each year. The nylon suture comes in either monofilament (e.g., Ethilon® and Dermalon®) or braided (e.g., Surgilon® and Nurolon®) forms. The braided sutures are more pliable and have better handling properties, but more costly than monofilament sutures. Nylon sutures have high mechanical strength, high elasticity, and low cost. It is also inert and, therefore,

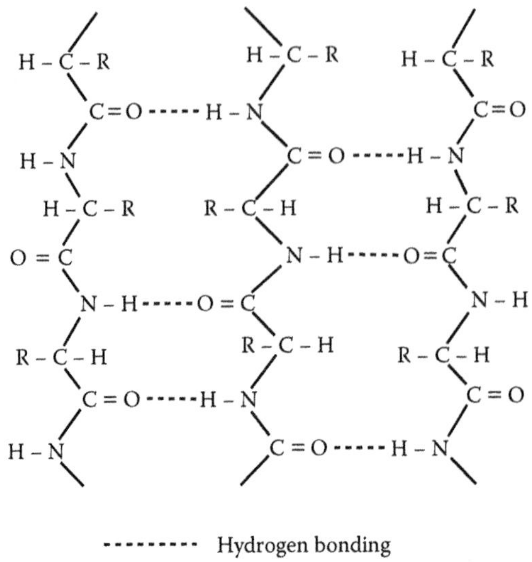

-------- Hydrogen bonding

FIGURE 6.5 The pleated β-sheet structure.

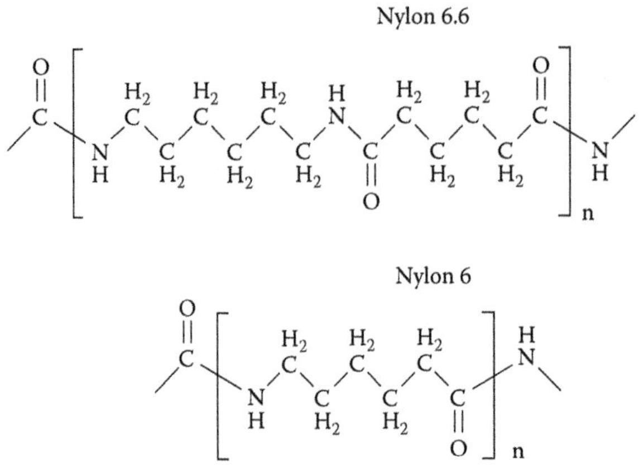

FIGURE 6.6 Chemical structure of nylon.

imposes minimal tissue responses. However, nylon has low knot security because of their tendency to return to original state (memory), and thus, an increased number of knot throws is needed to hold a stitch in place. An increase in the size or number of the knots, on the other hand, may lead to increased tissue responses [8].

"Polyester" suture is made from polyethylene terephthalate (PET), as shown in Figure 6.7. PET fibers have high mechanical strength, toughness, and abrasion resistance, as well as high elongation and elastic recovery. However, polyester fibers require higher forces to elongate as compared to nylon fibers. They also have higher resistance to heat and chemicals (including acids and oxidizing agents) than nylon fibers. In fabricating a PET suture, PET fiber is melt spun, stretched, and braided into

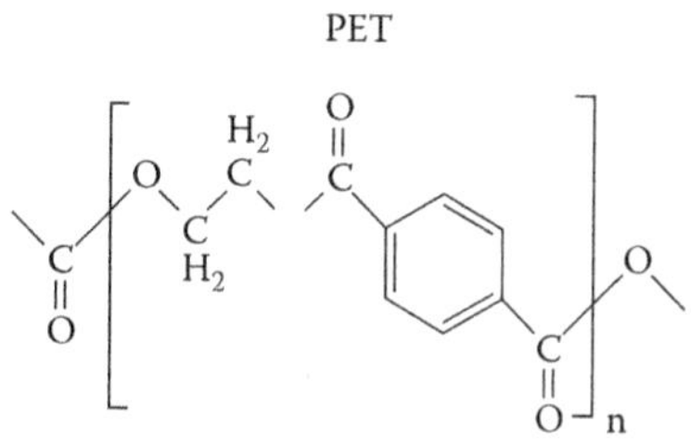

PET

FIGURE 6.7   Chemical structure of polyethylene terephthalate.

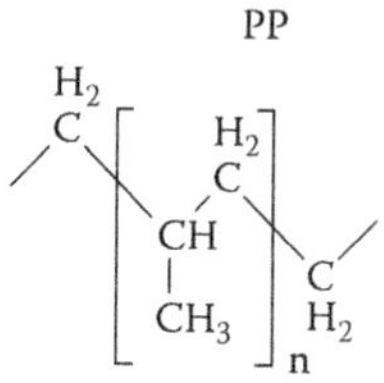

PP

FIGURE 6.8   Chemical structure of polypropylene.

a multifilament strand. It has lower tissue response than silk sutures. They can be uncoated (e.g., Mersilene® and Dacron®) or coated (e.g., Ethibond®). Uncoated sutures have a rougher surface and may produce drag when pulled through the tissue, and are therefore difficult to handle but have good knot security. Other polyester sutures are coated with polybutilate, silicone or Teflon® to increase their surface smoothness, at the cost of decreased knot security, however. The coatings may also enhance their tissue response. There have also been concerns about the coating's cracking after knots are tied [9]. Since polyester sutures maintain their high mechanical strength indefinitely after implantation, they can be used to fix implanted prosthetic materials [20].

"Polypropylene (PP)" is a linear hydrocarbon polymer, as shown in Figure 6.8. PP is known therefore for their high strength, toughness, and abrasion resistance. They have medium elasticity but high elastic recovery. PP fibers have the lowest capacity of water absorption among the commonly used fibers, and are thus called the zero absorbent fibers. PP suture (e.g., Surgilene® and Prolene®) is a flexible monofilament in form, with an above-the-average strength among sutures made of different materials. Polypropylene is also highly inert and thus causes very low tissue response or adherence to the tissue. A distinctive character of polypropylene suture is that it has an extremely smooth surface and exerts a minimal drag force when the suture is pulled through the tissue. The force required to remove the suture has been reported to be about one-third to two-thirds that of nylon and one-fourth to one-half that of silk [21]. As a result, polypropylene can be a good choice for cutaneous procedures because during its removal, it can be a "pull-out" suture with little disturbance to the tissue. The smooth surface, on the other hand, may cause low knot security as the knots tend to slip easily. Polypropylene suture is known for its good plasticity. It stretches and deforms to accommodate wound swelling to avoid cutting through the tissue.

"Polytetrafluoroethylene" (PTFE) is the most inert polymer of all, and therefore PTFE sutures (e.g., Cytoplast®) induce minimal tissue response. Expanded PTFE is used to produce monofilament sutures (ePTFE; e.g., Gore-Tex® suture) known for their high porosity. High porosity may reduce mechanical strength of the sutures, but give them a unique compression capacity. This feature allows a needle-to-suture diameter ratio of nearly 1, which contributes to reduced suture hole blood leakage as compared to other sutures without compressive ability. In addition, the PTFE sutures also impose minimal drag to the tissue during the implantation and removal procedures [22,23].

Sutures made from other materials are also known, but are not used as widely as the above-mentioned because of their performance:

1. Surgical cotton is a natural cellulose fiber; cotton suture can be handled easily but is physically weak. Its degradation rate is also difficult to predict. It usually loses half of its strength in 6–9 months.
2. Polyethylene (e.g., Dermalene®) suture is similar to polypropylene but has less tensile strength and knot security. It also slowly loses strength and eventually breaks down [20].
3. Polybutester (e.g., Novafil®) is a nonabsorbable suture material occurring in monofilament form. It has high elasticity and memory, similar to a rubber band [9].
4. Stainless steel is the strongest and least tissue reactive among all suture materials. However, it is stiff and may break at sites of bending, twisting, and knotting. Handling and knotting of steel sutures are difficult, but the knots are secure. They are usually used only for bone fixation [20].

Among the wide variety of sutures available today, not one of them possess performance attributes that would suit all types of end uses. The two major properties (strength and tissue reactivity) of the sutures are compared in Tables 6.2 and 6.3.

TABLE 6.2
Relative Tensile Strength of Commonly Used Suture Materials

Relative Tensile Strength	Nonabsorbable Sutures	Absorbable Sutures
High	Steel	
↑	Polyester	PGA
	Nylon (monofilament)	Polyglactin 910
	Nylon (braided)	
	Polypropylene	PDO
	Silk	
Low		Catgut

Source: Bennett, R.G., *J. Am. Acad. Dermatol.*, 18(4), 619–640, 1988.

TABLE 6.3
Relative Tissue Reactivity of Commonly Used Suture Materials

Relative Tissue Reactivity	Nonabsorbable Sutures	Absorbable Sutures
High		Catgut
↑	Silk, cotton	
	Coated polyester	Polyglactin 910
	Uncoated polyester	PGA
	Nylon	
	Polypropylene	
Low	Steel	

Source: Bennett, R.G., *J. Am. Acad. Dermatol.*, 18(4), 619–640, 1988.

In addition to the performance of a suture, its choice also depends on criteria such as cost and type of surgery. In some situations, there is the need for the temporary presence of a device or implant that does not have to be surgically removed when they are no longer needed. For example, sutures are needed to close the incision into a heart after an open heart surgery until the wound has healed. Degradable suture materials are useful for such applications. For cutaneous procedures (i.e., procedures on the surface of the skin), on the other hand, degradable suture materials do not appear to have advantages over nondegradable sutures materials in terms of both tensile strength and biocompatibility (as can be seen from Tables 6.2 and 6.3). Nondegradable sutures, therefore, seem to have a better fit for such applications.

6.2.2 Artificial Ligaments

Ligaments are strong and highly elastic strips of tissue that connect one bone to another. There are four main stabilizing ligaments at the knee joint (Figure 6.9): anterior cruciate ligament (ACL), posterior cruciate ligament (PCL), medial collateral ligament (MCL), and lateral collateral ligament (LCL). Among them, ACL is the one that connects the femur to the tibia and controls the extent to which the tibia can move in relation to the femur. ACL is the most frequently injured ligament and can be torn at a force of about 3.3 times the body weight [24]. ACL injuries are usually associated with sports that require sudden hyperextension or rotation forces, such as basketball, soccer, football, and skiing. Injured ACL has poor intrinsic healing potential and needs surgical intervention. It is estimated that in the United States about 95,000 new ACL injuries occur every year, and more than 50,000 ACL reconstructions are performed annually [25].

Three types of treatment are available. "Autograft" involves taking tendon tissues (i.e., a fibrous cord or band that connects a muscle with its bony attachment [26]) from the patient to have his ACL reconstructed. Tissues for this purpose are harvested from the hamstring (i.e., a tendon bounding the space behind the knee [26])

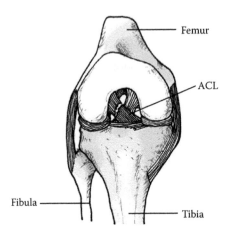

FIGURE 6.9 Anterior cruciate ligament at the knee joint.

and patella (i.e., knee cap) tendons [27]. Autograft is generally believed to be the best treatment because there is no issue of host responses. However, it has the drawback of additional injury and pain in the harvesting site, and can be limited by the availability or adequacy of healthy tissues from the patient. The next option is "allograft," in which graft tissues are obtained from a donor's cadaver. Allograft, however, is associated with such concerns as host responses, mechanical properties, and risk of disease transmission [28]. The last option for treating ACL injuries is the use of "ligament prosthesis." Synthetic materials such as PET, PTFE, and carbon fibers have been used to produce ligament prosthesis. A performing ligament prosthesis is expected to have a high tensile strength, a high elongation and recovery, and a stiffness that matches that of the host ACL. Braided structure is again the best choice for such performance requirements. For example, circular braids with an axial reinforcement of bundles of core yarns or braided structures have been developed and found to resemble the mechanical properties of a native ACL [29]. The choice of ligament prosthesis materials is also based on the mechanical requirements, as discussed below.

The "carbon fiber" is a high-performance inorganic fiber under continuous development in the last 50 years. It is composed of long and thin graphite sheets (shown in Figure 6.10) stacked together in bundles to form fibers. Currently there is a rich variety of carbon fibers available for different end uses, among which two are the most frequently used: the polyacrylonitrile (PAN)-based and pitch-based carbon fibers [30]. The PAN-based carbon fibers are processed from high-molecular-weight PAN via dry or wet spinning, and are the strongest carbon fibers. The pitch-based carbon fibers are subdivided into general-purpose and high-performance fibers. General-purpose pitch-based carbon fibers are prepared from high boiling fractions of petroleum feed stocks via melt spinning, while high-performance pitch-based carbon fibers are produced from mesophase (or known as the liquid crystal phase) pitch. Liquid crystals are highly structured and oriented liquids: their molecules flow like liquids but orient in a way as that of crystals [31]. As a result, high-performance

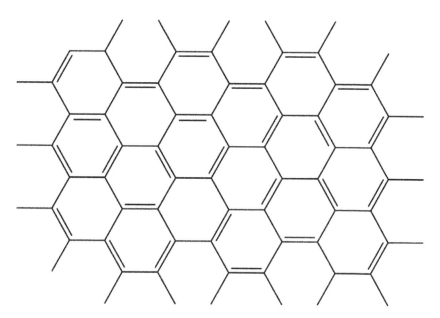

FIGURE 6.10 A small section of a graphite sheet.

pitch-based carbon fibers have high crystallinity and orientation, and therefore have even higher modulus than PAN-based carbon fibers. Because of their excellent mechanical properties, carbon fibers have been used in the aerospace industry, sports and recreational devices, and architecture in the form of reinforcements for composite materials. Also, carbon fibers have found applications in such implantable devices as ligament prosthesis, reinforcement for bone grafts.

Carbon fibers are used as ligament prosthesis in the form of braided bundles. They have high mechanical strength and good biocompatibility, and have been reported to support tissue proliferation [32]. However, the drawback of carbon fibers used as ligament prosthesis is their high fragility and brittleness, which will result in relatively low resistance to abrasion and strain failure [33].

PTFE ligament prostheses are fabricated from braided filaments of expanded PTFE (a porous material comprising about 75% air by volume), containing approximately 180 strands [34]. Short-term clinical outcomes have shown that the grafts have excellent biocompatibility, easy handling, and reduced healing time [35]. However, reports about problems related to PTFE ligament prosthesis increased in 1990s. The problems include graft breakage and lessening, and inflammation, largely due to the relatively low mechanical strength of PTFE. As a result, the PTFE graft was withdrawn from the market in 1993 [36].

PET is a popular material for ligament prosthesis due to its good biological and mechanical properties [37]. It is also available in the form of braided filament strands. A clinical follow-up study suggested that the long-term (1–8.3 years) stability of PET ligament prosthesis is acceptable, with over 80% good and excellent results (subjective patients' evaluation). The rest, however, may fail within a short time due to different causes (e.g., inflammation, ruptures, and infection), or lead to complaints like pain, swelling, and lack of motion [38]. The breakdown of PET in the body is through hydrolysis after long-term exposure to water/body fluid [39].

Ultrahigh-molecular-weight polyethylene (UHMWPE) is a relatively new material for ligament prosthesis. Therefore, clinical data for its performance are limited. In a 14-year prospective study on a small number of patients (16), the UHMWPE braided ligament prosthesis was found to function in a way similar to autograft. Such an UHMWPE braided ACL prosthesis contains more than 4000 filaments, and has a high tensile strength (9000 N) [40].

Other fiber materials that have been used to produce ligament prosthesis include polypropylene [41], aramid fibers [42], and carbon/PET fiber composites [43]. Despite decades of research and development work in the area, there has been no perfect graft for ACL reconstruction. There are problems to be solved. Long-term stability is still the major concern. Some of the major problems associated with current ligament prosthesis are discussed below.

Creep is the slow, gradual, and permanent deformation of a material due to the application of a constant stress, which may be below the yield strength of the material. At the yield point, a material undergoes permanent deformation, that is, when the stress is removed, the material will not return to its original shape. The stress corresponding to the yield point is called "yield strength." It is important in the design of an implant that its yield strength should be higher than the load level to which it will be subjected. Ligament prosthesis, for example, if subjected to a load higher than its yield strength, will undergo permanent deformation. As a result, its capacity to sustain further load will be significantly reduced, which may lead to failure of the implant.

Other mechanical problems include fatigue caused by cyclic loads and wear fragments as a result of abrasion damage. Clinical symptoms that follow may be pain, swelling, and inflammation, which can lead to the failure of a graft, too. There is also concern about the weak link where ligament prosthesis is surgically fixed to a bone, usually via staples or screws. Failures at these sites may be ascribed to the implantation procedures [38].

With such problems looming so large, usefulness of the ligament prosthesis for ligament reconstruction has been a controversial issue. Efforts to bring about significant improvement, however, have not been slackened nor suspended. Among these, tissue engineering seems to be promising, as will be discussed in Section 6.3.

6.3 PRODUCT DESIGNING AND EXAMPLES

Product designing of braided structures for applications in the medical fields, such as product designing in other fields, is a process involving a series of interrelated events and decisions, starting with knowing who will be the user or what will be the end use. The user or the end use will shape the concepts that ultimately dictate design and performance criteria. A number of novel designs have been reported in recent years for sutures and ligament prosthesis, as efforts to overcome different problems or challenges in their clinical applications. Three such designs are introduced in Sections 6.3.1 through 6.3.3.

6.3.1 BARBED SUTURE

The knot-tying process during the surgery leads to residual force and distortion of the tissue. The suture knots may obstruct blood perfusion through capillary vessels and hinder wound healing. They are also the source of significant tissue response. The question is: Is there any way to minimize knotting or remove the knot during the use of sutures?

When it is possible, a surgeon may choose to use a suture that will result in the smallest possible knots (e.g., a suture that is small in diameter) or/and require the smallest possible number of knots (e.g., a suture with better knot security). A newly developed barbed suture may provide a solution to this problem: such a suture is designed to be self-anchoring and knotless [44]. A first trial of a nonabsorbable barbed suture was conducted when a nylon barbed suture was applied in a flexor tendon repair, which proved to be an innovation that led to reduced tissue response [45]. However, the nonabsorbable suture needs to be removed afterward. To date, most of the barbed sutures are made from absorbable materials such as polydioxane, so that no removal procedure is needed. A bidirectional barbed suture (Figures 6.11 and 6.12) contains barbs in a spiral pattern around the circumference of the suture, with the barbs divided into two groups facing each other in opposing directions [46]. Clinical trials have suggested that the use of knotless barbed sutures can facilitate surgical wound closure [47].

FIGURE 6.11　A sketch of a bidirectional barbed suture.

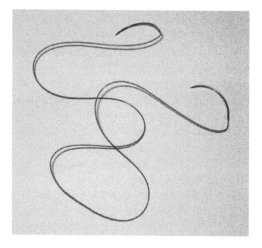

FIGURE 6.12　Bidirectional barbed suture. The device is composed of delayed-absorbable polydioxanone (PDO), cut with barbs that change direction at the mid-point of the double-armed suture. (From Siedhoff, M.T. et al., *J. Minim. Invasive Gynecol.*, 18(2), 218–223, 2011. With permission.) Copyright (2011) Elsevier.

6.3.2 Smart Sutures

Intelligence is a capacity of human beings or living creatures. The most admirable of such intelligence is man's ability to "sort of" "know" something simply according to some "feeling," without resorting to any extra, material means (a tool, for example), and to respond to that something (regarded now as a "stimulus"), by modifying one's behavior to effect positive outcomes. Over the decades, a variety of intelligent textiles and smart products have been used to promote health and quality of life. One such example is to engineer a suture material to avoid a current surgical problem, as discussed below.

There is a well-known challenge in endoscopic surgery: it is very difficult to use normal instruments to form a knot to close an incision with an appropriate stress. That is, if the knot is formed with a force that is too strong, it may cause damage to the surrounding tissue; if the force is too weak, there may be the danger of hernia due to scar tissue that is lower in mechanical strength than healthy tissue.

As a result, it is desirable that the suture is so smart that it can be temporarily elongated and applied loosely into the surgical position, that the temperature can be raised above the traditional point, and that the suture will duly shrink and tighten the knot to provide an optimum force. Biocompatible and biodegradable shape memory polymers (SMPs) were first introduced as a potential material for "intelligent sutures," and as a solution to the above-mentioned problem.

Shape memory materials are materials that have the capacity to "remember" their original shape; for example, they can return to their previous shape on exposure to a stimulus (e.g., deformation due to external stress). The most frequently used are thermally induced shape memory materials. Certain metal alloys and polymers also demonstrate such a smart capacity. A reversible change in the shape of a shape memory material results from a phase transition during the heating process at a certain temperature: for a shape memory alloy (e.g., the Nickel–Titanium alloy), reversible crystal transformation is involved; for an SMP, there usually occurs a reversible transformation from a glassy state to a rubbery state [48].

A suitable polymer material for this application can also be made to consist of a hard and a soft segment. The hard segments (e.g., polyurethane) form a crystalline region and determine the permanent shape. The soft segments (e.g., polyether or polyester diol), on the other hand, construct an amorphous region that is responsible for the deformation, retaining the temporary shape, and returning to the original shape on exposure to the predetermined stimulus. For example, a degradable multi-block polymer is reported to contain a hard segment of oligo(p-dioxanone)diol and a soft segment of oligo(e-caprolactone)diol, the two segments being coupled with 2,2(4),4-trimethylhexanediisocyanate. Recently, scientists have developed a suture of thermoplastic SMP, which can be prestretched for about 200% before forming a loose knot, and the knot could be tightened in 20 seconds when heated to 40°C. In vitro and animal model tests have validated its status as an "intelligent" suture [49]. This is a promising design of smart suture, which can be pre-elongated and "frozen" before the surgery. During the surgical procedures, the sutures can be applied relatively loosely into the body. After the surgery, the body temperature warms up the suture to "defrost" it, and as a result, the suture shrinks to tighten the knot with an optimum force.

6.3.3   BRAIDED SCAFFOLD FOR TISSUE ENGINEERED ACL

The injury or disease of a ligament, similar to the situations affecting other types of tissue or organ is a grievous and costly event in health care, affecting millions of people worldwide annually. Only a few types of tissues can be spontaneously regenerated (i.e., restoring the normal structure and function) when injured. These are epithelial tissues (e.g., epidermis), bone tissues, and smooth muscle tissues (i.e., one of the muscles of the internal organs, having no striations). Even for the regenerative tissue like bone, it is difficult or impossible to have their large defects repaired/restored. Those that do not have the capacity to regenerate themselves, including ligaments, need surgical intervention, of which the latest means is tissue engineering for the purpose of tissue regeneration.

Currently regarded as the most promising approach to treating damaged or failed human tissues/organs, tissue engineering refers to "an interdisciplinary field that applies the principles of engineering and the life sciences toward the development of biological substitutes that restore, maintain, or improve tissue function" [50]. In a tissue engineering process, appropriate living cells are seeded on a matrix or scaffold and then guided to develop into a new and functional living tissue. Grafts as a result of such regeneration are known as bioengineered grafts, which will be able to function as substitutes for injured or diseased skin, vascular, and bone tissues. Generally speaking, tissue engineering is based on the tripod of scaffold (i.e., environment for the living of cells), regulator(s) (such as growth factors), and cells (such as stem cells).

A scaffold is a 3D porous solid biomaterial that has been developed to perform some or all of the following functions, that is, to (1) provide an environment that is suitable for the cells to live and will promote cell adhesion and proliferation; (2) allow transport of gas, nutrients, and regulators that are necessary for the cells to perform their functions and to proliferate; (3) biodegrade at a controllable manner; and (4) cause minimal host responses. Depending on the specific tissue/organ to be regenerated, different biomaterials can be used for such a scaffold [5]. In the area of tissue-engineered ligament prosthesis, the scaffolds are usually in the form of braided structures [51,52].

Similarly, different application areas of tissue engineering usually require the use of different types of cells, each of which may demand a specific living environment. Creating appropriate environment for the survival and proliferation of one or even several types of cells has always been a challenge in the design of a tissue-engineered scaffold. One of the environmental factors that have been noticed and studied is the porosity of the scaffolds. When this rule is applied in ACL tissue engineering, for example, it has been shown that bone tissues prefer a scaffold with a minimum pore size of 150 μm for their ingrowth, whereas soft tissues require a scaffold with pore size in the range of 200–250 μm [53,54].

As a result, a desirable scaffold for tissue-engineered ACL is expected to provide different pore sizes at different locations: For bony attachment sites, where bone tissues will grow, it should be constructed to have higher density (lower porosity) and consequently higher mechanical strength to adapt to the rigors of surgical fixation onto the bone. On the other hand, the scaffold is expected to provide higher porosity to facilitate the growth of soft tissues. One reported design is to fabricate a braided structure with different density/porosity at different zones, as shown in Figure 6.13.

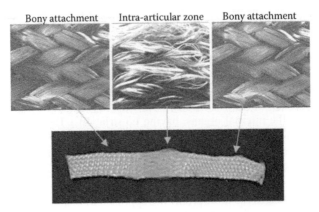

Bony attachment   Intra-articular zone   Bony attachment

4 × 12 3-D Braid PLAGA

FIGURE 6.13 A design of ligament scaffold for 3D rectangular braid. (From Cooper, J.A. et al., *Biomaterials*, 26(13), 1523–1532, 2005. With permission.) Copyright (2005) Elsevier.

The femoral and tibial tunnel attachment sites have high angle fiber orientation and, as a result, higher density (or lower porosity), while the intra-articular zone has lower angle fiber orientation, and thus a lower density (higher porosity) [55]. Different fiber types have also been evaluated for such a design, and the results suggested that poly-L-lactic acid may be a more suitable material for ACL regeneration as compared to PLGA and PGA [56]. Both in vitro and in vivo evaluations demonstrate that this could be a promising design in the ACL tissue regeneration [55,56]. Some other development in this area involves the use of appropriate growth factors to induce the cell proliferation and tissue regenerations [57,58].

6.4 SUMMARY

This chapter provides a discussion of the various braided structures to be used biologically, that is, used as a material to be implanted into the human flesh to be in contact with human tissues. Such use predetermines the many prerequisites for the use, especially that the material and device to be implanted must biologically suit the microclimate in the human body, and will stay there for a relatively long period, which are what we call biocompatibility and biostability, the two fundamental properties of any biomaterial or biological device, as highlighted throughout the chapter.

The discussion of materials is performance oriented. Since performance is the combined consequence of various properties, the discussion of performance is a discussion of properties. This discussion of properties includes both desirable properties and undesirable ones, so that we know what material is compatible with its end use. For instance, the performance of sutures depends on physical properties such as elasticity and knot strength, which are closely related to their handling properties such as pliability and knottability, which further indicate how easily the suture can be handled by the surgeon, and also depends on the extent to which the use of sutures cause tissue response, an obviously undesirable property of suture materials. A clarification of both desirable and undesirable properties favors wise choice of materials.

This discussion of properties is important because it is related to specific structures or materials (sometimes with product brands) and devices. For the same purpose, many materials have been listed for sutures: catgut, polyamide (nylon), PDO, polyester, polyglactin 910, PGA, polypropylene, PTFE, and silk. And discussions of materials are often aided with tables, where they can be directly related to properties.

The braided structures are known for their excellent mechanical properties, including tensile strength and flexibility, along the longitudinal direction and therefore have been applied in products that desire such performance. For their applications as implants, biocompatibility is another critical property for the materials to be chosen. The future trend will be continuing on optimizing the medical products in terms of both biocompatibility and biostability.

REFERENCES

1. Sumanasinghe, R.D. and M.W. King, New trends in biotextiles: The challenge of tissue engineering. *JTATM*, 2003. **3**(2): 1–13.
2. Mak, T.W. and M.E. Saunders, *The Immune Response: Basic and Clinical Principles.* 2006, Boston, MA: Elsevier/Academic. xx, 1194 p.
3. Park, J.B. and R.S. Lakes, *Biomaterials: An Introduction*, 2007, Springer: New York, NY. p. xi, 561 p.
4. Ratner, B.D. et al., Biomaterials Science. In *An Introduction to Materials in Medicine*, 2004, Boston, MA: Elsevier Academic Press. p. xii, 851 p.
5. Lanza, R.P., R.S. Langer, and J. Vacanti, *Principles of Tissue Engineering*. 3rd ed. 2007, San Diego, CA: Academic Press. xli, 995 p.
6. Bennett, R.G., Continuing medical education: Selection of wound closure materials. *J Am Acad Dermatol*, 1988. **18**(4): 619–640.
7. Greenberg, J.A. and R.M. Clark, Advances in suture material for obstetric and gynecologic surgery. *Rev Obstet Gynecol*, 2009. **2**(3): 146–58.
8. Meyer, R.D. and C.J. Antonini, A review of suture materials, Part I. *Compendium*, 1989. **10**(5): 260–2, 264–5.
9. Moy, R.L., B. Waldman, and D.W. Hein, A review of sutures and suturing techniques. *J Dermatol Surg Oncol*, 1992. **18**(9): 785–795.
10. Levin, M.P., Periodontal suture materials and surgical dressings. *Dental Clin N Am*, 1980. **24**(4): 767–781.
11. Morgan, M.N., New synthetic absorbable suture material. *Br Med J*, 1969. **2**(5652): 308.
12. Herrmann, J.B., R.J. Kelly, and G.A. Higgins, Polyglycolic acid sutures: Laboratory and clinical evaluation of a new absorbable suture material. *Arch Surg*, 1970. **100**(4): 486–490.
13. Craig, P.H. et al., Biologic comparison of polyglactin-910 and polyglycolic acid synthetic absorbable sutures. *Surg Gynec Obstet*, 1975. **141**(1): 1–10.
14. Barnes, C.P. et al., Nanofiber technology: Designing the next generation of tissue engineering scaffolds. *Adv Drug Delivery Rev*, 2007. **59**(14): 1413–1433.
15. Rodeheaver, G.T. et al., A temporary nontoxic lubricant for a synthetic absorbable suture. *Surg Gynec Obstet*, 1987. **164**(1): 17–21.
16. Ray, J.A. et al., Polydioxanone (PDS), a novel mono-filament synthetic absorbable suture. *Surg Gynec Obstet*, 1981. **153**(4): 497–507.
17. Hatch, K.L., *Textile Science*. 1993, Minneapolis/Saint Paul, MN: West Pub., 472 p.
18. Postlethwait, R.W., D.A. Willigan, and A.W. Ulin, Human tissue reaction to sutures. *Ann Surg*, 1975. **181**(2): 144–150.
19. Altman, G.H. et al., Silk-based biomaterials. *Biomaterials*, 2003. **24**(3): 401–416.

20. Meyer, R.D. and C.J. Antonini, A review of suture materials, Part II. *Compendium*, 1989. **10**(6): 360–2, 364, 366–8.
21. Freeman, B.S. et al., An analysis of suture withdrawal stress. *Surg Gynecol Obstet Internl Abst Surg*, 1970. **131**(3): 441–448.
22. Ratner, B.D., *Biomaterials Science: An Introduction to Materials in Medicine*. 2nd ed. 2004, Boston, MA: Elsevier Academic Press. xii, 851 p.
23. Gayle, R.G. et al., Evaluation of the expanded polytetrafluoroethylene (EPTFE) suture in peripheral vascular-surgery using EPTFE prosthetic vascular grafts. *J Cardiovasc Surg*, 1988. **29**(5): 556–559.
24. Noyes, F.R. and E.S. Grood, The strength of the anterior cruciate ligament in humans and Rhesus monkeys. *J Bone Joint Surg Am*, 1976. **58**(8): 1074–82.
25. Frank, C.B. and D.W. Jackson, The science of reconstruction of the anterior cruciate ligament. *J Bone Joint Surg Am*, 1997. **79**(10): 1556–76.
26. Dirckx, J.H., *Stedman's Concise Medical Dictionary for the Health Professions: Illustrated*. 4th ed. 2001, Baltimore, MD: Lippincott Williams & Wilkins.
27. Spindler, K.P., R.A. Magnussen, and J.L. Carey, does autograft choice determine intermediate-term outcome of ACL reconstruction? *Knee Surg Sports Traumatol Arthrosc*, 2011. **19**(3): 462–472.
28. Nutton, R.W., I. McLean, and E. Melville, Tendon allografts in knee ligament surgery. *J R Coll Surg Edinb*, 1999. **44**(4): 236–240.
29. Cruz, J. et al., Designing artificial anterior cruciate ligaments based on novel fibrous structures. *Fiber Polym*, 2014. **15**(1): 181–186.
30. Hearle, J.W.S., *High-Performance Fibres*. 2001, Boca Raton, FL: CRC Press/Woodhead. p. xi, 329 p.
31. Khoo, I.-C., *Liquid Crystals*. 2nd ed. Wiley series in pure and applied optics. 2007, Hoboken, NJ: Wiley-Interscience. xiv, 368 p.
32. Osborne, A.H., R.C. Telfer, and M.A. Farquharson-Roberts, Recent experience with carbon fibre. *J Roy Nav Med Serv*, 1984. **70**(2): 66–69.
33. Blazewicz, S., C. Wajler, and J. Chlopek, Static and dynamic fatigue properties of carbon ligament prosthesis. *J Biomed Mater Res*, 1996. **32**(2): 215–219.
34. Indelicato, P.A., M.S. Pascale, and M.O. Huegel, Early experience with the Gore-Tex polytetrafluoroethylene anterior cruciate ligament prosthesis. *Am J Sports Med*, 1989. **17**(1): 55–62.
35. Vogel, U.B., Clinical-experiences with the Gore-Tex 5-ptfe cruciate-ligament prosthesis: Greater-than-4-year follow-up. *Helv Chir Acta*, 1992. **58**(6): 943–947.
36. Muren, O. et al., Gross osteolytic tibia tunnel widening with the use of Gore-Tex anterior cruciate ligament prosthesis: A radiological, arthrometric and clinical evaluation of 17 patients 13–15 years after surgery. *Acta Orthop*, 2005. **76**(2): 270–274.
37. Seitz, H. et al., Biocompatibility of polyethylene terephthalate (Trevira® hochfest) augmentation device in repair of the anterior cruciate ligament. *Biomaterials*, 1998. **19**(1–3): 189–196.
38. Krudwig, W.K., Anterior cruciate ligament reconstruction using an alloplastic ligament of polyethylene terephthalate (PET-Trevira®-hochfest). Follow-up study. *Bio-Med Mater Eng*, 2002. **12**(1): 59–67.
39. Hamid, S.H., M.B. Amin, and A.G. Maadhah, *Handbook of Polymer Degradation. Environmental Science and Pollution Control Series*. 1992, New York, NY: M. Dekker. x, 649 p.
40. Purchase, R. et al., Fourteen-year prospective results of a high-density polyethylene prosthetic anterior cruciate ligament reconstruction. *J Long-Term Effects Med Implants*, 2007. **17**(1): 13–9.

41. Kdolsky, R.K. et al., Braided polypropylene augmentation device in reconstructive surgery of the anterior cruciate ligament: Long-term clinical performance of 594 patients and short-term arthroscopic results, failure analysis by scanning electron microscopy, and synovial histomorphology. *J Orthop Res*, 1997. **15**(1): 1–10.
42. Dauner, M. et al., Para-aramid fiber for artificial ligament. In *Clinical Implant Materials*, edited by G. Heimke, U. Soltesz, and A.J.C. Lee. Vol. 9. 1990, Amsterdam, The Netherlands: Elsevier Science Publ B V. 445–449.
43. Campbell, A.C. and P.S. Rae, Anterior cruciate reconstruction with the abc carbon and polyester prosthetic ligament. *Ann Roy Coll Surg Eng*, 1995. 77(5): 349–350.
44. Siedhoff, M.T., A.C. Yunker, and J.F. Steege, Decreased incidence of vaginal cuff dehiscence after laparoscopic closure with bidirectional barbed suture. *J Minim Invasive Gynecol*, 2011. **18**(2): 218–223.
45. McKenzie, A.R., An experimental multiple barbed suture for the long flexor tendons of the palm and fingers. Preliminary report. *J Bone Joint Surg Br*, 1967. **49**(3): 440–447.
46. Dattilo, P.P. et al., Medical Textiles: application of an absorbable barbed bidirectional surgial suture. *JTATM*, 2002. **2**(2): 1–5.
47. Einarsson, J.I. et al., Use of bidirectional barbed suture in laparoscopic myomectomy: evaluation of perioperative outcomes, safety, and efficacy. *J Minim Invasive Gynecol*, 2011. **18**(1): 92–95.
48. Langenhove, L.v., editor. *Smart Textiles for Medicine and Healthcare: Materials, Systems and Applications*. 2007, Cambridge, United Kingdom: Woodhead Publishing. xiii, 312 p.
49. Lendlein, A. and R. Langer, Biodegradable, elastic shape-memory polymers for potential biomedical applications. *Science*, 2002. **296**(5573): 1673–1676.
50. Langer, R. and J.P. Vacanti, Tissue engineering. *Science*, 1993. **260**(5110): 920–926.
51. Araque-Monros, M.C. et al., New concept for a regenerative and resorbable prosthesis for tendon and ligament: Physicochemical and biological characterization of PLA-braided biomaterial. *J Biomed Mater Res Part A*, 2013. **101**(11): 3228–3237.
52. Barber, J.G. et al., Braided nanofibrous scaffold for tendon and ligament tissue engineering. *Tissue Eng Part A*, 2013. **19**(11–12): 1265–1274.
53. Spector, M. et al., High-modulus polymer for porous orthopedic implants: Biomechanical compatibility of porous implants. *J Biomed Mater Res*, 1978. **12**(5): 665–677.
54. Konikoff, J.J. et al., Development of a single stage active tendon prosthesis.1. Distal end attachment. *J Bone Joint Surg*, 1974. **A 56**(4): 848–848.
55. Cooper, J.A. et al., Fiber-based tissue-engineered scaffold for ligament replacement: Design considerations and in vitro evaluation. *Biomaterials*, 2005. **26**(13): 1523–1532.
56. Lu, H.H. et al., Anterior cruciate ligament regeneration using braided biodegradable scaffolds: In vitro optimization studies. *Biomaterials*, 2005. **26**(23): 4805–4816.
57. Jenner, J.M.G.T. et al., Effect of transforming growth factor-beta and growth differentiation factor-5 on proliferation and matrix production by human bone marrow stromal cells cultured on braided poly lactic-co-glycolic acid scaffolds for ligament tissue engineering. *Tissue Eng*, 2007. **13**(7): 1573–1582.
58. Kimura, Y. et al., Regeneration of anterior cruciate ligament by biodegradable scaffold combined with local controlled release of basic fibroblast growth factor and collagen wrapping. *Tissue Eng*, 2008. **14**(1): 47–57.

7 Applications of Braided Structures in Civil Engineering

Fernando Cunha, Daniel Oliveira,
Graça Vasconcelos, and Raul Fangueiro

CONTENTS

ABSTRACT

This chapter presents an overview of the applications of braided structures in civil engineering. A brief introduction is provided on fiber reinforced plastics (FRPs), different fibers, and polymeric resins used in civil construction. More specifically, this chapter discusses the utilization of braided structures in concrete reinforcement and masonry wall strengthening. Particular focus has been given on the recently developed braided composite rods (BCRs), and for each of these applications the technique for producing BCRs or BCR-based materials, application methods, and obtained results are discussed in detail. The comparison of performance of braided structures with that of conventional FRP-based materials is also presented in this chapter.

7.1 INTRODUCTION

In his search for better civil engineering solutions, man has been trying to mix materials to obtain innovative materials with enhanced performance. Taking advantage of particular properties of each component, it has been possible to create synergies able to transform the construction industry. A remarkable example of such changes is steel reinforced concrete. By combining concrete and steel in a single structural element, two different properties were set together, compressive and tensile strengths, resulting in such a huge advancement that even a century after its development it is still the most used structural material (Waldron 2004; Karbhar 2007).

Although steel reinforced concrete is an excellent combination of materials, a big problem that affects its durability and lifetime expectation is corrosion of steel. Corrosion of steel takes place when the thin protective oxide layer produced during the hydration of cement is destroyed. The destruction of this layer can occur essentially in two ways: corrosion induced by carbonation, which occurs due to the carbon dioxide present in the air, and corrosion by chlorides, near seawater environments or when very high levels of chlorides are present. When the corrosion propagates, different levels of damage can happen, affecting the performance of the structure and consequently its loading capacity and safety (see Figure 7.1) (Moreno et al. 2004; Bertolini et al. 2004; Waldron 2004).

Several methods have been proposed to solve this problem of steel corrosion, such as increasing concrete cover over steel reinforcements, use of epoxy-coated steel bars, application of coatings on steel bars, use of corrosion inhibitors in concrete, cathodic protection, and so on (Waldron 2004). A proposed solution that captured the attention of the most forward-thinking engineers was the application of fiber reinforced polymer composites (FRP) as reinforcing elements in concrete. FRPs are composite materials that were developed in the 1950s by the aerospace industry and consist of high-strength and high-stiffness fibers combined with a high-performance

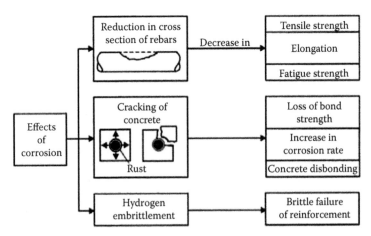

FIGURE 7.1 Structural consequences of corrosion in reinforced concrete structures. (From Bertolini et al., Corrosion of Steel in Concrete: Prevention, Diagnosis, Repair, John Wiley & Sons, Weinheim, Germany, 2004.)

thermosetting polymer, containing other fillers or additives to achieve other properties if necessary. In an FRP, the fibers are the reinforcement, responsible for carrying loads in the direction of the fibers, whereas the polymer provides protection to the fibers and transfers the stress from fiber to fiber (Marques 2011; Bank 2006; Karbhar 2007). The combination of these two (or more) materials created new materials with undeniable advantages: high strength to self-weight ratio (10–15 times greater than steel), low axial coefficient of thermal expansion, excellent fatigue characteristics, corrosion resistance, potentially high durability, and electromagnetic transparency.

Currently, a wide variety of fibers that can be used exists. It is usual to classify fibers according to their origin. Fibers can be natural, such as jute, sisal, and coconut among others, or produced by human being. Within the latter group, fibers can be divided into organic fibers, such as Kevlar, and inorganic fibers, such as glass, carbon, and basalt (Figure 7.2) (International Bureau for the Standardisation of Man-Made Fibers 2006).

The most common fibers used in FRP applications are glass (G), carbon (C), aramid (A), and more recently basalt (B); see Figure 7.3 for a graphical comparison.

The most commonly used thermoset resins in FRP materials for civil engineering applications are polyester resin (unsaturated), vinylester resin, and epoxy resin (Bank 2006; Balaguru, Nanni, and Giancaspro 2008). The polyester resin is mainly used for producing industrial and commercial FRP composites due to its low price and easy manipulation. This resin is also used for producing structural profiles and some FRP rods; but when greater protection to external agents, greater extension at

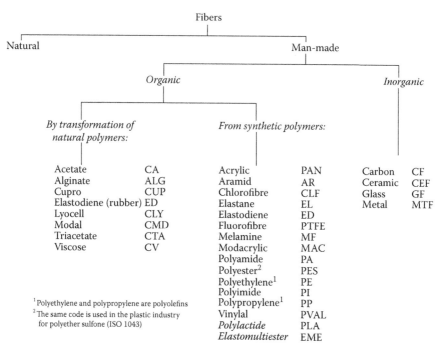

FIGURE 7.2 Fiber classification. (From International Bureau for the Standardisation of Man-Made Fibers, *Terminology of Man-Made Fibers*, BISFA, Brussels, Belgium, 2006.)

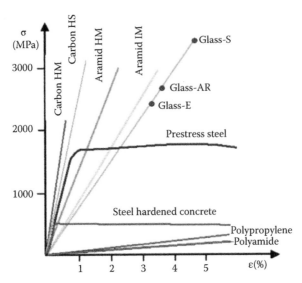

FIGURE 7.3 Mechanical properties of most common fibers and metallic materials. (From Marques, P., "Design of Fibrous Structures for Civil Engineering Applications" (master's thesis), University of Minho, Guimarães, Portugal, 2009.)

break, and lower drying shrinkage is necessary, vinylester or epoxy resins are preferred. Vinylester resin is the most used resin to produce FRP rods and profiles due to its good corrosion resistance, ease of processing, and high resistance to alkaline environment. These characteristics are due to the fact that this polymer is a hybrid of epoxy and polyester resins. Finally, the epoxy resin is the most used resin in FRP products for structural reinforcement, being used in most precured FRP systems, FRP-support adhesives in laminates, and others. Its main features are great corrosion resistance, low drying shrinkage, high crack resistance when exposed to high temperatures, and resistance to high temperatures. The major disadvantage is the cost of the resin, as shown in Table 7.1.

Another great advantage of FRP materials is their different shape possibilities, essentially due to the large spectrum of existing production techniques: from hand layup to vacuum bagging system, pultrusion, filament winding, resin transfer molding, injection molding, compression molding, and braiding. Therefore, it has been possible to produce a great variety of FRP shapes such as bars, grids, sheets, plates, and profiles (Bank 2006; Waldron 2004; Balaguru, Nanni, and Giancaspro 2008; Hota et al. 2007).

Since the 1960s, corrosion in existing steel reinforced concrete structures started to be a visible problem, especially in important structures; but it was only in the 1970s and 1980s that the study of FRP bars gained importance to avoid the problem of reinforcement corrosion. At the time, some pioneering work using small-diameter FRP bars (6 mm) to reinforce or to prestress concrete structural members was carried out, and some companies such as Marshall-Vega and International Grating, Inc., started to develop the first glass fiber reinforced polymer (GFRP) rods for concrete reinforcement applications. At that time, these materials were

TABLE 7.1

Mechanical Properties of Thermoset Resins

	Specific Weight (mg/m³)	Young's Modulus (GPa)	Tensile Strength (MPa)	Notched Impact Strength (kJ/m²)	Fracture Toughness (MPa·m$^{1/2}$)	Coeff. of Thermal Expansion (10$^{-6.}$°C^{-1})	T_{max} Service (°C)	Rupture Strain (%)	Postcure Shrinkage (%)
Unsaturated polyester	1.1–1.5	2–4.5	40–60	1–2.5	0.5–1.7	100–180	90–130	2–2.5	5–15
Vinylester	1.1–1.2	3.5–4	65–85			40–70	100–150	3–6	8–9
Epoxy	1.1–1.4	2–5	50–100	1–8	0.5–2	50–120	120–180	2–8	0.1–2.5
Phenolic	1.2–1.5	2.5–4.5	30–60	1–2	0.8–1.5	115–130	140–200	1–3	0.1–1.5

Source: Fangueiro, R., *Fibrous and Composite Materials for Civil Engineering Applictions*, Woodhead Publishing, Ltd., Cambridge, United Kingdom, 2011.

used to build magnetic resonance imaging facilities, due to their electromagnetic transparency (Bank 2006; ACI Committee 440 2002).

By the end of the 1980s, the reinforcement corrosion problem induced another boost in the research on FRP materials, and the composites industry came up with more developments in FRP bars produced through the pultrusion process. At this time, glass fiber was the most used fiber and the resins used were vinylester and polyester, and to enhance the bond between FRP reinforcing bars and concrete different strategies were applied: sand coating the FRP, deforming the FRP surface, and FRP bars with helically wound spiral outer surfaces. In the beginning of the 1990s, carbon fiber was already introduced, and it was applied to carbon fiber reinforced polymer (CFRP) rods, leading to relevant research about the behavior of concrete reinforced with FRP bars. The industry of FRP bars was blooming in the United States, Japan, and Europe (Benmoktane, Chaallalt, and Masmoudi 1995; Bank 2006; ACI Committee 440 2002). A drawback of concrete sections reinforced with FRP realized at that time was its less ductile and fragile response (Figure 7.4), due to the FRP's elastic behavior without plastic deformation (Benmoktane, Chaallalt, and Masmoudi 1995).

By the end of the 1990s, the FRP bar world market had grown up, with different types of internal reinforcements being available such as isopod bars from Pultrall, a pultruded bar composed of E-glass fiber and a polyester resin (Figure 7.5). Its natural smooth pultruded surface was improved by helical winding of glass fiber around the bar and by coating sand particles of a specific grain-size distribution (Pultrall, Inc.; Bank 2006).

A two-dimensional grid product, called new fiber composite material for reinforced concrete (NEF-MAC), was developed by Shimizu Corporation, Japan. This grid-type material was available in glass, carbon, and hybrid glass/carbon fiber types (Figure 7.6) (ACI Committee 440 2002; Karbhari 1998).

Carbon Fiber Composite Cable (CFCC™) (Figure 7.7), a stranded cable, was produced by Tokyo Rope and was composed of 7, 19, or 37 twisted carbon fibers. The objective of this material was to use it for pretensioning and internal or external post-tensioning in civil structures (ACI Committee 440 2002; Karbhari 1998).

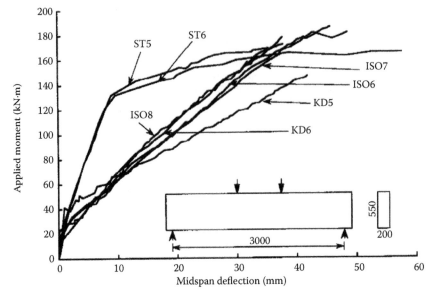

FIGURE 7.4 Load–deflection curve of concrete beams reinforced with steel (ST5 and ST6) and with glass fiber reinforced polymer bars (ISO6, ISO7, ISO8, KD5, and KD6). (From Benmoktane et al., *Construction and Building Materials*, 9(6), 353–364, 1995.)

FIGURE 7.5 Pultruded fiber reinforced plastic bars. (From Pultrall, Inc., *Pultrall*, Publi-Web.net. n.d. http://www.pultrall.com/vrod.asp?languageid = 2, accessed January 5, 2013.)

Leadline™ (Figure 7.8), another type of CFRP prestressing tendon, used for prestressing (pre- and posttensioning), was produced by Mitsubishi Chemical. This material was available in different diameters with smooth surface bars (for smaller diameters) and with deformed (ribbed or indented) surfaces (ACI Committee 440 2002; Karbhari 1998).

FiBRA™, an aramid fiber–based material, was produced by the braiding process and impregnated by an epoxy resin. The braiding process made it possible to efficiently manufacture large-diameter bars, between 3 and 20 mm, and provided a natural deformed surface configuration, improving the mechanical interlocking with concrete, as shown in Figure 7.9. Fiber rods are still available in the market in the form of rigid rods or flexible cables (Nanni et al. 1993; ACI Committee 440 2002).

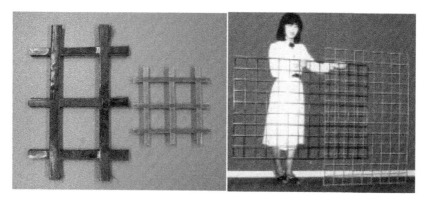

FIGURE 7.6 NEF-MAC reinforcing grids. (From Bank, L. C., *Composites for Construction—Structural Design with FRP Materials*, John Wiley & Sons, Inc., New Jersey, 2006; Karbhari, V. M., *Use of Composite Materials in Civil Infrastructure in Japan*, International Technology Research Institute—World Technology (WTEC) Division, Maryland, 1998.)

CFCC U   CFCC1 × 7   CFCC1 × 19   CFCC1 × 37

FIGURE 7.7 Examples of Carbon Fiber Composite Cable for prestressing concrete. (From Karbhari, V.M., *Use of Composite Materials in Civil Infrastructure in Japan*, International Technology Research Institute—World Technology (WTEC) Division, Maryland, 1998; Bank, L.C., *Composites for Construction—Structural Design with FRP Materials*, John Wiley & Sons, Inc., New Jersey, 2006.)

FIGURE 7.8 Japanese fiber reinforced plastic prestressing tendons. (From Bank, L. C., *Composites for Construction—Structural Design with FRP Materials*, John Wiley & Sons, Inc., New Jersey, 2006.)

Technora™ is an FRP bar manufactured by Sumitomo Construction and Teijin (textile industry), and it is made by pultrusion and consists of straight aramid fibers impregnated with vinylester resin, with a spirally wound yarn to improve the bonding with concrete (Figure 7.10). This material was available even for prestressing application by a special bond type anchorage system (ACI Committee 440 2002; Karbhari 1998).

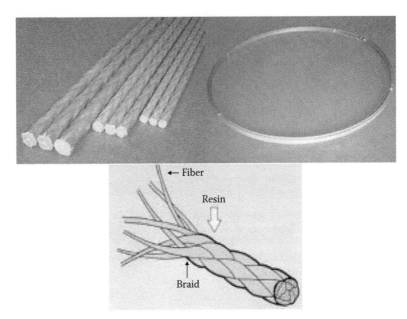

FIGURE 7.9   Rods and cables produce by Fibex. (From Nanni et al., *Cement Concrete Comp.*, 15, 121–129, 1993; ACI Committee 440. *State-of-the-Art Report on Fiber Reinforced Plastic (FRP) Reinforcement for Concrete Structures—ACI 440R-96*. American Concrete Institute, Michigan, USA, 2002.)

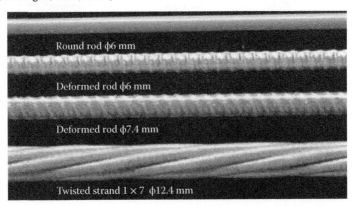

FIGURE 7.10   Examples of some Technora fiber reinforced plastic rods. (From Karbhari, V.M., *Use of Composite Materials in Civil Infrastructure in Japan*, International Technology Research Institute—World Technology (WTEC) Division, Maryland, 1998.)

To regulate the designing of concrete structures reinforced with FRP bars, different design codes have been developed, which are presented in chronological order as follows (Reis 2009):

- Japan Society of Civil Engineers: Recommendation for Design and Construction of Concrete Structures Using Continuous Fibre Reinforcing Materials (Japan, 1997)

- Intelligent Sensing for Innovative Structures (ISIS): ISIS Design Manual No. 3: Reinforcing Concrete Structures with Fibre Reinforced Polymers (Canada, 2001)
- American Concrete Institute (ACI)—440.3R-04: Guide Test Methods for Fibre-Reinforced Polymers (FRPs) for Reinforcing or Strengthening Concrete Structures (USA, 2004)
- American Concrete Institute (ACI)—440.1R-06: Guide for the Design and Construction of Structural Concrete Reinforced with FRP Bars (USA, 2006)
- Federation International de Beton (Fib)—fib bulletin n°40: FRP Reinforcement in RC Structures (provisional version) (Europe, 2007)
- Italian National Research Council (CNR)—CNR-DT 203/2006: Guide for the Design and Construction of Concrete Structures Reinforced with Fibre-Reinforced Polymer Bars (Italy, 2007)

Although extensive research has been carried out in the last few years and existing products have been developed and design codes prepared, some problems have still not been solved. The rods composed of a single fiber type have a linear elastic behavior and fragile rupture, without displaying any warning or ductility before rupture. Therefore, design theories and FRP design codes enforce the use of high safety coefficients to prevent fragile rupture of the reinforcements and, instead, to obtain rupture of the concrete section through the ductile crushing of the compressed lamina of concrete. Therefore, bigger reinforcing ratios are needed and the FRP's high strength is not fully explored (Balendran et al. 2002; Federation International de Beton 2007). Also, the stiffness of standard FRP bars is much lower compared to steel; in glass fiber rods it is approximately one-eighth of steel's value, and in standard carbon fibers it is half of steel's value. This causes big difficulties in structural applications (Cui et al. 2008; Wu 2005). There are still doubts about the durability of FRP bars inside concrete, especially in the case of GFRP rods composed of glass fibers of type E that are degraded in alkaline environments. Other durability aspects are the absorption of moisture in aqueous environments, ice–deice effect, high temperatures, ultraviolet radiation, and fire resistance (Micelli and Nanni 2004; Ceroni, Gaetano, and Pecce 2006; Wang, Wong, and Kodur 2007).

7.2 CONCRETE REINFORCEMENT

Steel corrosion demands rehabilitation procedures for all kind of constructions. In concrete reinforced structures, corrosion causes the cracking of concrete. In aggressive environments, such as seaside constructions, this problem becomes much more significant, turning the preservation, repair, and rehabilitation of structures into a periodic need. Moreover, the weight of reinforcing materials is a factor of paramount importance, namely, in logistic costs due to transportation to the working place, handling and placing in situ and due to the limitation on the production and reinforcement of lightweight reinforced concrete elements.

7.2.1 Introduction

Fiber reinforced plastic (FRP) rods are currently proposed as reinforcing bars for reinforced concrete. This type of rebar is designated to provide a solution to the problem of durability of ordinary steel bars, which do not always exhibit adequate corrosion resistance in severe environments despite new protection techniques being available. FRP rods exhibit good mechanical properties as well as corrosion resistance in concrete structures subjected to severe conditions (Katz 2000). Figure 7.11 illustrates different groups of FRP rods as follows: GFRP, CFRP, basalt fiber reinforced polymer, aramid fiber reinforced polymer, and a new generation of rods called BCRs.

FRP rods are generally manufactured by pultrusion, where the longitudinal fibers are drawn through a resin bath and then passed through a die, which gives the rod its final shape. Unless additional treatment is given, a smooth surface is obtained on the rods, which prevents good adhesion, or bonding, to the surrounding concrete (Katz 1999).

7.2.2 Braided Composite Rods

BCRs are core reinforced tubular braided structures impregnated by a polymeric matrix. BCRs are produced by the braiding technique, a conventional textile technique with the ability to replace steel in the reinforcement of concrete elements. These BCRs present tensile strengths higher than those of conventional steel rebars. BCRs are corrosion free; therefore, repair and rehabilitation procedures due to steel corrosion are not required. This product is also much lighter than steel rebars (83%). Thus, it diminishes the logistic costs due to transport to the working place, as well as handling and placing in situ. Moreover, BCRs allow the reinforcement of lightweight concrete elements where the reinforcing material weight is an essential factor.

7.2.3 Production Process

The braiding process employs mutual intertwining of yarns. BCRs were recently developed in a modified braiding machine developed at the University of Minho, Portugal (Figure 7.12). The manufacture of composites has limitations based on shape, microstructure, and materials. Theoretically, any combination of these three aspects should be possible, although the development is a complex process and requires simultaneous consideration of various parameters such as production volume, reinforcement type, matrix type, and reinforcement and matrix relative volumes.

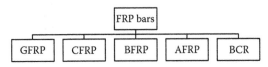

FIGURE 7.11 Typology of fiber reinforced plastic bars. AFRP, aramid fiber reinforced polymer; BFRP, basalt fiber reinforced polymer.

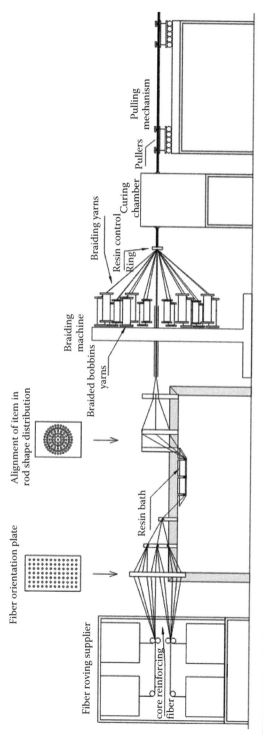

FIGURE 7.12 Braiding process.

The production process is started by the selection of fibrous materials to introduce in the composite rods. This stage has extreme importance in the process as it is the stage in which the mechanical behavior of the material is designed. Hence, the fibers are preselected according to their mechanical behavior, as shown in Table 7.2.

The most important property of the fibers is their elastic modulus, because they must be significantly stiffer than the matrix to carry the stress applied to the composite material. Therefore, fibers must also have sufficient strength to avoid failure. It is noted that fibers present a linear elastic response up to the ultimate load, with no significant yielding (Fangueiro and Gonilho-Pereira 2011).

After choosing the fiber types and quantities, they are taken in the fiber roving supplier and oriented through a fiber orientation plate (Figure 7.12). Once oriented, the fibers are passed to the resin bath. There is a wide choice of resins available, many of which are suitable for forming composites. Some restrictions are placed on the use of polyester resins for composite materials. Nevertheless, the choice depends on the required durability, the manufacturing process, and the cost. Generally, thermosetting resins are used. Table 7.3 illustrates the different types of thermosetting resins.

After fiber impregnation according to the nature of the chosen resin, fibers pass through an alignment segment to orient the fibers in the correct position and remove the excess resin. Afterward, the fibers are introduced as a core of the braided structure. The braided structure is composed of an external layer made of polyester fibers. These polyester fibers are selected to enhance the properties in terms of protection of

TABLE 7.2
Typical Fiber Properties

Fiberglass	Tensile Strength (N/mm²)	Modulus of Elasticity (kN/mm²)	Elongation (%)	Specific Density
E	2.54	70	2200	3–3.5
R	2.5	86	3200	3.5–4
S	2.53	86	3500	4.1
A	2.7	75	1700	
Aramidic				
Kevlar® 29	1.44	59	2640	3–5
Kevlar® 49	1.45	130	2900	1.5–3
Kevlar® 149	1.47	146	2410	1–2
Carbon				
High strength	1.80	230	4500	2.0
Intermediate modulus	1.76	290	3100	1.1
High modulus	1.86	380	2700	0.7
Ultra high modulus	1.94	588	3920	0.7
Basalt	4840	89	3.15	2.7

Source: Fangueiro, R., and C. Gonilho-Pereira, *Fibrous and Composite Materials for Civil Engineering,* Woodhead Publishing, Ltd., Cambridge, United Kingdom, 2011; Marques, P., "Design of Fibrous Structures for Civil Engineering Applications" (master's thesis), University of Minho, Guimarães, Portugal, 2009.

TABLE 7.3
Typical Resin Properties

Resin	Tensile Strength (N/mm²)	Modulus of Elasticity (kN/mm²)	Elongation (%)
Polyester	55–70	3.6–4.1	1.5–3.0
Vinylester	68–80	3.5	3.5–6
Epoxy	65–120	3.0–4.1	2.0–8.0

Source: Fangueiro, R., and C. Gonilho-Pereira, *Fibrous and Composite Materials for Civil Engineering,* Woodhead Publishing, Ltd., Cambridge, United Kingdom, 2011.

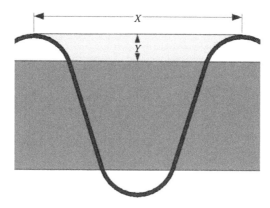

FIGURE 7.13 Adherence square.

core fibers and adherence between concrete and braided structure. These fibers also perform an important role in keeping the alignment of the core fibers.

The surface of braided rods can be tailored to give better compatibility with the concrete. The distance between the braided yarns can be modified (parameter X) and also the diameter of yarns (Y) (Figure 7.13). The distance between the braiding yarns is controlled by modifying the takeoff speed. Also, the takeoff speed modifies the braiding angle and there is a negative correlation between the takeoff speed and the braiding angle.

After the braiding stage, the material is placed in the curing chamber, which promotes the curing of the thermoset resin. The curing speed is approximately 0.5 m/min, with a temperature of 180°C.

7.2.4 MECHANICAL BEHAVIOR

Gonilho and Fangueiro (Fangueiro and Gonilho-Pereira 2011) produced seven different BCRs using polyester fibers for the braided structure, E-glass, carbon and HT polyethylene fibers for the core, and a polyester resin in the resin bath for the impregnation of the core. All BCRs were produced with the same braided structure geometry and linear density, only varying the type of core reinforcement fiber, as shown in Table 7.4. Different combinations were evaluated for the composition of the core, using a single type of fibers as well as a blend of two or three different types of fibers and varying the percentage of each one. The objective was to evaluate the influence of the fibers and its percentage on the mechanical behavior of BCRs. Table 7.4 presents the percentage of fibers used as core reinforcement over the total linear density (mass per unit length) of the core reinforcement. The mechanical behavior of the rods produced was evaluated according to ASTM D 3916-94, with a crosshead speed of 5 mm/min. Table 7.5 presents the values of the tensile test results obtained for each type of rod. It is clear that the mechanical properties of BCR strongly depends on the type of core and therefore, it is possible to tailor the mechanical properties of BCR varying the core fibre type and composition.

TABLE 7.4
BCR Compositions

| Rod Type | Type of Core Reinforcement Fiber | | |
	E-Glass Fiber (%)	Carbon Fiber (%)	HT Polyethylene Fiber (%)
1	100	0	0
2	77	23	0
3	53	47	0
4	0	100	0
5	50	45	5
6	52	45	3
7	75	22	3

Source: Fangueiro, R., and C. Gonilho-Pereira, *Fibrous and Composite Materials for Civil Engineering*, Woodhead Publishing, Ltd., Cambridge, United Kingdom, 2011.

TABLE 7.5
Tensile Test Results Obtained for the Different BCRs

Rod Type	Tensile Strength (MPa)	Extension at Failure	Tensile Strength at 0.2% (MPa)	Modulus of Elasticity (GPa)
1	485	0.01701	110	55
2	766	0.01416	157	78
3	740	0.01178	148	74
4	747	0.01183	192	96
5	679	0.01105	167	83
6	652	0.01098	162	81
7	690	0.01438	146	73

Source: Fangueiro, R., and C. Gonilho-Pereira, *Fibrous and Composite Materials for Civil Engineering*, Woodhead Publishing, Ltd., Cambridge, United Kingdom, 2011.

7.2.5 Self-Monitoring Composite Rods

Monitoring allows the analysis of the behavior along the different phases of construction, usage, and hazardous situations. In this context, the main motivation for the development and application of monitoring systems was to improve the safety of the structures, as well as economic considerations.

Different methods/techniques are applied in several situations. Vibrational monitoring is a nondestructive method that is based on the evaluation of response of a string/rod to a vibration stimulus. This technique requires that the monitored materials should not be confined within other materials. Optical behavior monitoring is a well-studied subject, with solutions existing in the market. This method allows measurement of tensile stress, extension, temperature, and instant acceleration. However, it presents substantial costs, which make it unaffordable for standard applications in conventional constructions. Finally, some piezoresistive materials allow the combination of reinforcement properties with piezoelectric characteristics. The piezoelectric materials have the capability to change their electrical resistance or conductivity according to their mechanical load and deformation in a proportional manner (electrical resistance and mechanical load or deformation). In this way, it is possible to use a single material for performing simultaneously as reinforcing and monitoring elements.

University of Minho has been doing several studies on this last monitoring technique, by studying the capability of BCRs for reinforcement and monitoring of civil constructions (Fangueiro and Gonilho-Pereira 2011). For this purpose, hybrid, carbon fiber (CF), and glass fiber (GF) reinforced composite rods were tested for monitoring performance. It was reported by other studies that the memorizing ability of composite materials resulted from the increased residual resistance of the composites after loading–unloading cycles, which was in relation with the previously applied maximum strain. Under prestressed conditions, the composite rods showed remarkable memorizing ability with lower detectable strains. A main drawback raised by these studies was related to damage detection: damage cannot be detected at early stages, unless carbon fibers are replaced with carbon particles. However, in the latter case mass production was more expensive and demanding (Fangueiro and Gonilho-Pereira 2011).

In a hybrid composite rod, carbon fiber and the relative percentages of carbon fiber and glass fiber are the factors that determine the variations of electrical resistance with stress and/or strain. As mentioned earlier, the main drawback of hybrid CF–GF reinforced polymer composites is the lack of sensitivity at lower strains, which can be improved by prestressing (Fangueiro and Gonilho-Pereira 2011).

Fibrous materials Research Group (Fangueiro and Gonilho-Pereira 2011; Fangueiro et al. 2013; Rosado Mérida et al. 2012; Rana et al. 2014) performed research on the development of self-monitoring pseudoductile BCRs based on a conventional braiding machine with minor modifications, developed at the University of Minho. Three types of BCRs with different carbon fiber contents (23%, 77%, and 100%) were produced. Table 7.6 shows the BCRs' composition, whereas Figure 7.14 shows the corresponding composite cross sections.

To evaluate the piezoresistive concept, a cyclic loading test was performed on the three developed rods in which the mechanical and electrical properties were simultaneously evaluated. Figure 7.15 presents the increase in electrical resistance with

TABLE 7.6
Self-sensing BCR Compositions

Rod Type	Fiber Composition	Fiber (%)	No. of Rovings	Linear Mass (tex)
2	E-glass/carbon	77/23	18/3	1600/900
3	E-glass/carbon	53/47	53/47	900
4	Carbon	100	12	900

Source: Fangueiro, R., and C. Gonilho-Pereira, *Fibrous and Composite Materials for Civil Engineering*, Woodhead Publishing, Ltd., Cambridge, United Kingdom, 2011.

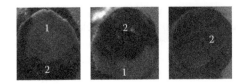

FIGURE 7.14 Braided composite rod cross sections for rod types 2, 3, and 4: 1, glass fiber; 2, carbon fiber. (From Fangueiro, R., and C. Gonilho-Pereira, *Fibrous and Composite Materials for Civil Engineering*, Woodhead Publishing, Ltd., Cambridge, United Kingdom, 2011.)

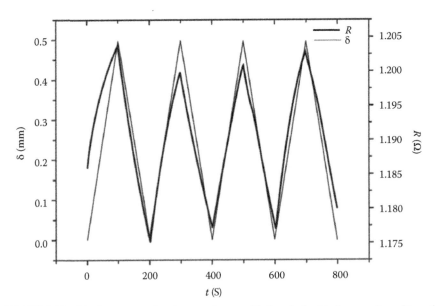

FIGURE 7.15 Displacement and resistance change with time for braided composite rod type 4 (100% carbon). (From Fangueiro, R., and C. Gonilho-Pereira, *Fibrous and Composite Materials for Civil Engineering*, Woodhead Publishing, Ltd., Cambridge, United Kingdom, 2011.)

increasing displacement. In general, the electrical resistance during loading and unloading increases linearly at lower displacement values and nonlinearly at higher ones. The tested sample showed the change of electrical resistance in proper compliance with the change of its deformation. The main factor influencing the type of response of each sample was the relative position of the fibers or, more precisely, the carbon fiber placement along the length of the rods (Fangueiro and Gonilho-Pereira 2011).

7.2.6 Ductility of Braided Composite Rods

FRP has been historically identified as an ideal candidate to address the aging problem of infrastructures (ACI 440R-96). In spite of its exciting advantages, the adaptation of BCR as concrete reinforcement has been slow for a variety of reasons, including high cost and lack of familiarity with a new technology. Another impediment to quick adaptation is the lack of ductility of currently available FRP systems. The tensile stress–strain behavior of current FRP reinforcement is essentially linear elastic up to the ultimate load followed by a brittle failure. Thus, the tensile behavior of these bars in flexure members shows very little ductility in the traditional sense.

In the late 1990s, Somboonsong, Ko, and Harris (1998) in the United States developed a kind of reinforced braided structure and worked with a ductile hybrid Kevlar–carbon reinforced plastic (D-H-FRP) (Figure 7.16) produced by an innovative technique called the "braidtrusion" process. This research was developed to solve the three main drawbacks of common FRPs: lack of ductility, low stiffness, and low bonding properties to concrete. The objective was to develop a ductile or pseudoductile rod that would meet the steel reinforcing bars' properties: high initial modulus, definitive yield point, and a high level of ultimate strain. The final objective was to design the material to directly substitute steel taking all advantages from FRP: high strength, low weight, and corrosion resistance (Somboonsong, Ko, and Harris 1998).

This hybrid braided composite rebar was composed of an external braided structure and core reinforcement. The braidtrusion process consists of a 24-carrier

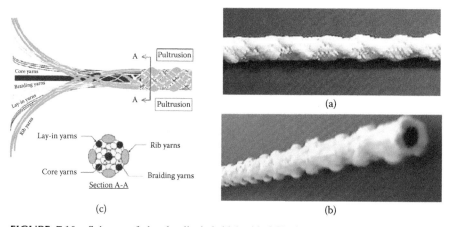

FIGURE 7.16 Scheme of the ductile hybrid braided Kevlar–carbon plastic (a) and (b). Finished product of 10 mm–diameter D-H-FRP braided rebar (c) (From Somboonsong et al., *ACI Mater. J.*, 95(6), 655–666, 1998.)

braiding machine that forms a tubular braided structure and allows the application of different types of yarns in the sheath structure and in the core. The braiding process occurs in the forming ring. Afterward, the tubular braided structure passes through an infusion zone where an epoxy resin impregnates the fibers. Then, the rod passes through a heated chamber to cure the resin and after 30 minutes the braided rod is ready to be collected (Pastore and Ko 1999).

With this type of three-dimensional textile structure and with a correct choice of fiber materials, it is possible to control, within limits, the load–deformation curve. In this case, aramid (Kevlar) fibers were applied in the braided structure to achieve a pseudoductile behavior, and high-modulus carbon fibers were applied in the core to solve the stiffness problems of common FRP bars. This structure also allows the creation of a rib texture on the surface by adding coarser yarns, creating in this way an improved interlocking with concrete, as shown in Figure 7.17 (Pastore and Ko 1999).

To design the material, a complex mathematical model that incorporates the effect of the hybrid material system was developed. The objective was to reach the failure of the Kevlar fibers present in the braided structure, in order to achieve a progressive failure that fits the requirement of 3% strain at failure. The design methodology considered the effects of fibers in different levels: from fiber level to yarn level, to weave level, and to braid level (Figure 7.18) (Somboonsong, Ko, and Harris 1998).

To validate the model, prototype samples of 3, 5, and 10 mm of diameter were produced containing high-modulus carbon fibers (Thornel P-55S) in the core and aramid (Kevlar 49) in the braided structure. Tensile tests have been performed, and the results, as illustrated in Figure 7.19, showed that the D-H-FRP prototype had a yield stress of 275 MPa, an ultimate strength of 406 MPa, and a Young's modulus of 78.6 GPa, considering the full area of cross section (Harris, Somboonsong, and Ko 1998). When only the core reinforcing area is considered, the Young's modulus goes up to 202 GPa, same as that of steel reinforcement (Hampton, Ko, and Harris 1999). The obtained curves showed a bilinear stress–strain tensile curve with a well-defined yield, an ultimate strength higher than yield, and an ultimate failure strain between 2% and 3% (Hampton, Ko, and Harris 1999). Besides, the pseudoductile behavior showed a good agreement with the developed theoretical model predictions, although the analytical model predicted a higher drop in stress at yield that did not really happen (Somboonsong, Ko, and Harris 1998).

A comparative test in concrete beams with a cross section of 50 mm by 100 mm and a length of 1.2 m was also done (Harris, Somboonsong, and Ko 1998).

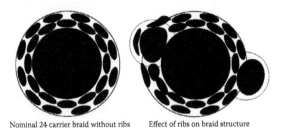

Nominal 24 carrier braid without ribs   Effect of ribs on braid structure

FIGURE 7.17   Effect of rib yarns on braided fabric geometry. (From Pastore, C.M., and F.K. Ko, *Braided Hybrid Composites for Bridge Repair*, National Textile Center Annual Report, Philadelphia and Drexel University, 1999.)

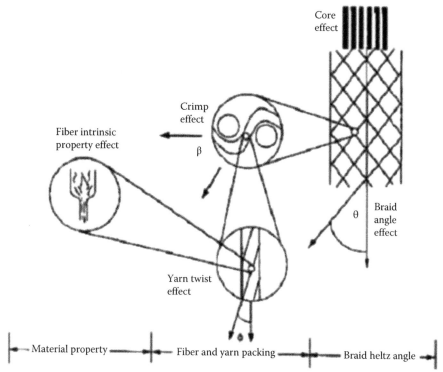

FIGURE 7.18 Structural hierarchy of fibrous assemblies: 1, fiber level; 2, yarn level; 3, weave level; 4, fabric/braid level. (From Somboonsong et al., *Aci. Mater. J.*, 95(6), 655–666, 1998.)

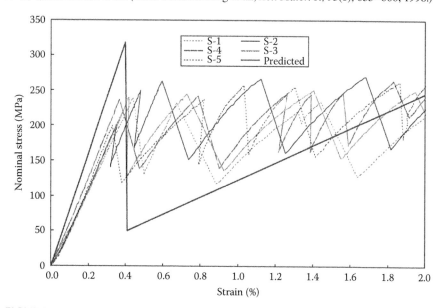

FIGURE 7.19 Monotonic stress–strain curves for 5-mm D-H-FRP bars. (From Somboonsong et al., *ACI Mater. J.*, 95(6), 655–666, 1998.)

Four beams were tested in four-point bending, three reinforced with D-H-FRP and one with steel. The results showed that the postcracking behavior of all three D-H-FRP reinforced beams had a well-defined bilinear load–deflection curve up to the yield point. After the yield point, the loading capacity continued to increase until it reached its ultimate strength, as shown in Figure 7.20. The loading–unloading cycles applied to each beam also showed the energy-absorbing capabilities and the reproducibility of the ductile behavior. It is worthwhile to note that the plastic deformation observed could work as a warning signal for the structure.

The performance in earthquake situations was also measured in small beams that represented portions of rigid frames (Figure 7.21) (Lam et al. 2001). These small beams were subjected to an increasing cyclic horizontal load that simulated the effect of an earthquake (Lam et al. 2001; Hampton, Ko, and Harris 1999; Harris et al. 2000).

The measured hysteretic behavior demonstrated the considerable energy dissipation capabilities of D-H-FRP (Harris et al. 2000), as shown in Figure 7.22. This

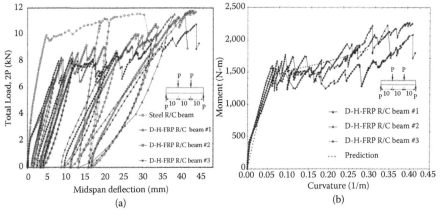

FIGURE 7.20 Comparison of (a) load–deflection behavior of steel and D-H-FRP reinforced beams and (b) the moment versus curvature. (From Harris et al., *J. Compos. Constr.*, 2, 28–37, 1998.)

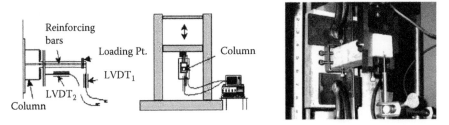

FIGURE 7.21 Cyclic loading test setup. LVDT, linear variable differential transducer. (From Lam et al., "Design Methodology of a Ductile Hybrid Kevlar-Carbon Reinforced Plastic for Concrete Structures by the Braidtrusion Process," 13th International Conference on Composite Materials, Beijing, China, 2001; Hampton et al., "Creep, Stress, Rupture, and Behavior of a Ductile Hybrid FRP for Concrete Structures," 12th International Conference on Composite Materials, Paris, France, 1999.)

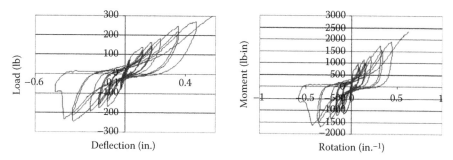

FIGURE 7.22 Load–deflection curve (left) and moment–rotation (right) for beam specimen under reverse cyclic loading. (From Harris et al., "Cyclic Behavior of a Second Generation Ductile Hybrid Fiber Reinforced Polymer (D-H-FRP) for Earthquake Resistant Concrete Structures," 12th World Conference on Earthquake Engineering (12WCEE), Auckland, Australia, 2000.)

energy dissipation capacity during seismic events is critical for structures, and it is not typical of FRP materials (Hampton and Ko 2007).

7.3 MASONRY WALL STRENGTHENING

The last seismic events in southern Europe have highlighted the vulnerability of the most usual constructive typology in contemporary architecture, that is, reinforced concrete framed structures with masonry infills (Lourenço et al. 2010). The nonstructural elements, such as masonry infills, have shown high vulnerability to seismic action, exhibiting a high degree of damage even for medium-intensity earthquakes, causing human casualties and high economic losses. For decades, these elements have been considered nonstructural and, therefore, they were not expected to play a structural role, even locally (Al-Chaar, Issa, and Sweeney 2002).

Therefore, there is a large segment within the building stock in earthquake-prone areas that needs to undergo preventive actions, particularly for out-of-plane loads. This can range from a simple connection of the masonry infills to the frame structures, which can also be applied in the case of already damaged elements, to the strengthening of the masonry infills. The potential benefits go beyond the stability of nonstructural elements, as this can improve the behavior of the whole structure in facing seismic events.

Laminated fiber reinforced polymer (FRP) strengthening of structures is a viable retrofitting technique due to its small thickness, advantageous strength to weight ratio, high stiffness, and relative ease of application. However, the application of these materials also has disadvantages like inadequate bond of the reinforcement to the masonry, resulting in the detachment of the reinforcement from the wall. Therefore, the alternative retrofitting technique that has been studied is based on fibers embedded in a matrix of mortar, a technique commonly referred to as textile reinforced mortar (TRM), and it may provide several advantages among which overcoming of bonding and problems with humidity are very important. Furthermore, it is possible to optimize the textile meshes inserted within masonry mortar, thereby improving the tensile strength of walls and deformation capacity, resulting in a more ductile failure. Following the use of TRM as a retrofitting technique for masonry infill walls, some new fiber-based materials for structural reinforcement based on braiding techniques have been recently developed at the Fibrous Materials Research Group (Fibrenamics), University of Minho, as an alternative to conventional FRP strengthening materials (Gómez 2012; Cunha 2012; Martins 2013). These materials have several advantages such as the possibility of designing the composition according to mechanical requirements and a low-technology and low-cost production process. Moreover, the shape of the external surface can be optimized through the variation of braiding process parameters, to get the best bonding behavior with the involved mortar. Therefore, one of the primary objectives of the past scientific activities carried out at the Fibrous Materials Research Group (Fibrenamics) was to optimize the external surface of BCR to enhance the adherence between mortar and BCR so that BCR meshes can be successfully applied for the retrofitting of brick masonry under flexural loads. The performance of these meshes has also been compared with the behavior of different commercial meshes available in the market.

7.3.1 MANUFACTURE OF BRAIDED RODS AND MESHES

Braided rods are manufactured through a technique also used for the production of fibrous reinforcements for construction applications (Fangueiro 2011), as already discussed in Section 1.3. It has been used for two centuries and is being increasingly used for technical applications. The braided rods for masonry walls are also produced from the combination of three types of materials, each with different functions, and manufactured through braiding of yarns in the transverse and longitudinal directions, forming a tubular structure. The yarns that form the base of the braided structure consist of a central core, which is responsible for the mechanical performance. Braiding yarns are taken from two groups of spindles that rotate in opposite directions, clockwise and counterclockwise. With the aim of improving mechanical properties and for adding new functionalities, axial fibers are added to the core of the braided rod (Figure 7.23a). In Figure 7.23b, the representative scheme of a cross section of BCR is provided. It is noted that 16 multifilaments of polyester are used for braiding, and multifilaments of glass or carbon fibers are used in the core. On the other hand, to fill the voids between the materials providing stability and homogeneity to the composite, a resin is applied on the external polyester surface.

Braiding angle is the most important parameter of the textile braided structure, influencing directly its mechanical behavior (Martins 2013). Braiding angle is the angle between the longitudinal axis and the direction of insertion of braiding yarns, as shown in Figure 7.24. The diameter of the braid is the straight line connecting the two extremities and passing through the braid center (see Figure 7.24). This parameter can vary according to the braiding yarn diameter, diameter of the axial structures, and circulation velocity (Martins 2013).

The final step of the preparation of the rods involves tensioning them with a force of 100 N and then stabilizing them. The tensioning was done by attaching the braided rods to a stable structure at one end and hanging a weight at the other end. Once tensioned, impregnation with resin is performed manually. The polyester resin is chosen because

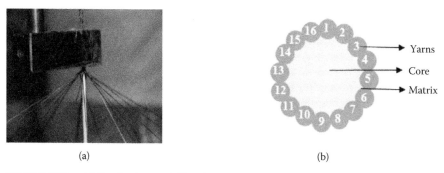

<table>
</table>

(a)   (b)

FIGURE 7.23   (a) Production and (b) schematic representation of the simple rod without roughness.

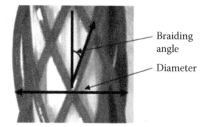

FIGURE 7.24   Braiding angle and diameter of a braided composite rod.

of its lower cost and toxicity compared to epoxy resins. It is activated through addition of 2% of methyl ethyl ketone peroxide. This technique aims at ensuring the bonding and homogeneity of both shell and core, as it is intended that they should work together and are stable enough for production of meshes by combining individual BCRs.

The produced rods are then assembled into a bidirectional mesh. A woven structure for the mesh is preferred to the simple superposition of rods in perpendicular directions, due to better self-stability of the mesh before its implementation and better guaranteed bidirectional behavior. The configuration of the connections of rods causes some roughness to the mesh, which can result in an additional imbrication (Figure 7.25). Different spaces can be used depending on the performance to be achieved in both directions.

7.3.2 Mechanical Characterization of the Braided Rods

The typical behavior of rods with glass and carbon fibers used in the core under uniaxial tension is characterized through uniaxial tensile tests according to NP EN ISO 2062 and ASTM 5034 standards (Figure 7.26). Based on the results, it is possible to conclude that the core reinforcement is responsible for BCR's initial stiffness and maximum load achieved. The initial stiffness and maximum force of carbon rods are higher than the rods with glass reinforcement. At the time of rupture, the polyester braided structure is totally loaded, which justifies the existence of a ductile plateau in both types of tested rods.

FIGURE 7.25   Details of mesh.

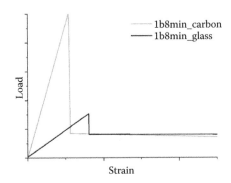

FIGURE 7.26   Typical load–strain response of different types of composite rods.

7.3.3   CHARACTERIZATION OF ADHERENCE OF BRAIDED RODS WITH MORTAR

One important requirement that needs to be fulfilled in considering braided rods as a reinforcing material for masonry walls (TRM) is the appropriate adherence between rods and the embedded mortar. The composite rods can be composed of different materials and different configurations of yarns, resulting in different roughnesses of external surface. On the other hand, there should be a surface roughness that optimizes the mechanical adherence of the rods to the mortar. To assess the influence of surface roughness of rods on adherence, different rods were produced by varying the structural parameters. For this purpose, polyester braids (one or two), composed of 8 or 16 polyester yarns, were added to the braided structure replacing simple polyester yarns and the structures were produced with different speeds, resulting in a total of 14 types of braided structures, as shown in Table 7.7.

For the adherence tests, cylindrical specimens of mortar (50-mm diameter and 150-mm height) were casted with a textile braided structure located at the center and embedded in the total height of the cylinder (Figure 27a). The mortar presented an average compressive strength of 3.56 MPa and an average flexural strength of 1.61 MPa. The specimens were kept in the laboratory conditions for 28 days. In the free tip of the rod, an extra length was kept to allow the connection and fastening of the rod to the test machine, in order to pull it out from the mortar specimen (Figure 27a). The linear

TABLE 7.7

Description of the Samples Used for Adherence Tests

Sample Designation	Description of the Samples	Speed (m/min)
0bmin	1 to 16 position: multifilament polyester with 11 tex	0.54
0bmax		1.07
1b8max	1 Position—braided multifilament consisting of 8 polyester with 11 tex 2 to 16 position: simple multifilament polyester with 11 tex	1.07
1b8int		0.8
1b8min		0.54
2b8max	1 and 9 position—braided multifilament consisting of 8 polyester with 11 tex 2 to 8 and 10 to 16 position: multifilament polyester with 11 tex	1.07
2b8int		0.8
2b8min		0.54
1b16max	1 Position: braided multifilament consisting of 16 polyester with 11 tex 2 to 16 position: multifilament polyester with 11 tex	1.07
1b16int		0.8
1b16min		0.54
2b16max	1 and 9 position: braided multifilament consisting of 16 polyester with 11 tex 2 to 8 and 10 to 16 position: multifilament polyester with 11 tex	1.07
2b16int		0.8
2b16min		0.54

Source: Martins, A., "Seismic strengthening solutions for masonry infill walls [Soluções de Reforço Sísmico de Paredes de Alvenaria de Enchimento]," University of Minho, Civil Department, Guimarães, Portugal, 2013.

sliding of the rod from the mortar cylinder was measured by a linear variable differential transducer. The cylindrical specimen of mortar is confined vertically through two steel sheets to promote the relative displacement between the rod and the mortar. The test speed was 0.010 mm/s, which corresponds to the test duration of approximately 45–60 minutes, suitable for this kind of test (Figure 7.27b).

The typical loading versus sliding curves resulting from the pull out tests are presented in Figure 7.28 for all produced braided structures. It is seen that the roughness formed by a simple braid with 8 yarns exhibits a better behavior than the roughness with 16 yarns, because they present better ductile behavior and adherence forces. The factors that influence the maximum adhesion force depend on the interaction of

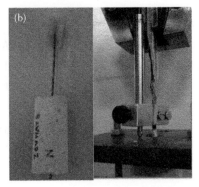

FIGURE 7.27 (a) Production of specimens for pull out tests and (b) test setup. (From Martins, A., "Seismic strengthening solutions for masonry infill walls [Soluções de Reforço Sísmico de Paredes de Alvenaria de Enchimento]," University of Minho, Civil Department, Guimarães, Portugal, 2013.)

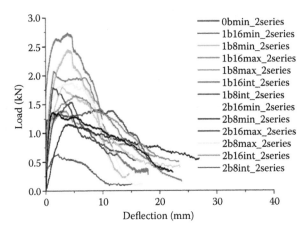

FIGURE 7.28 Results of pull out tests. (From Martins, A., "Seismic strengthening solutions for masonry infill walls [Soluções de Reforço Sísmico de Paredes de Alvenaria de Enchimento]," University of Minho, Civil Department, Guimarães, Portugal, 2013.)

the adherence area of the rods and the effect of interlocking controlled by the roughness. The most satisfactory performance in terms of maximum force was exhibited by the structure, 1b8max.

7.3.4 CHARACTERIZATION OF ADHERENCE OF BRAIDED ROD MESHES

Complementary to the pull out tests of the rods on mortar cylinders, it is also considered important to evaluate the adherence between the meshes and the mortar layer, as this is more representative of the real situation. For this, representative meshes were developed with a selected braided structure, taking into account the results of the pull out tests, to understand their behavior when they are pulled out inside the mortar plaster layer of masonry samples. The rods were constituted by two multifilaments of carbon with

1600 tex, similar to the previous tests, and the selected braided structure was 1b8min. Besides this braided structure, the original configuration without roughness (0bmin) was also considered to analyze the influence of roughness on adherence to mortar.

Beyond the study on the meshes constituted by BCRs, two commercial solutions with similar mechanical and physical characteristics were also tested to analyze the viability of produced meshes. These meshes are different in terms of type of fibers; the commercial mesh named as Comm_carb is constituted by carbon fiber in main direction and the commercial mesh named as Comm_glass is constituted by glass fibers. The commercial mesh composed of carbon fibers (Comm_carb) is unidirectional, taking into account that carbon fibers are oriented in the direction where bending occurs, and its density is 200 g/m with a spacing of approximately 25 mm in the main direction. The commercial mesh of glass fibers, Comm_glass consists of resistant glass fibers in both directions. Once bidirectional, the mesh density is 225g/m^2, with a spacing between fibers of 25 mm.

The construction of representative masonry specimens was made by an experienced mason to reproduce similar workmanship used in current structures. The samples were kept under relatively stable conditions of temperature and humidity inside the laboratory (Figure 7.29). Specimens were made by applying a thin layer of mortar, with subsequent placement of the reinforcing mesh embedded in a new layer of rendering mortar, giving a total thickness of about 20 mm. The mortar used in these tests was the same as the one used in the previous study. The dimension of inserted mesh in plastering was 200 mm × 100 mm (length × width), presenting the same free area on the samples. On the free tip of each mesh applied to the samples, a bond was created, allowing the connection and fastening of the mesh to the machine, similar to that created in individual rods.

These tests were controlled by displacement, and measurements were done by the internal system. The speed was 0.08 mm/s, by using the test setup shown in Figure 7.30. The specimen was fixed to the base of the frame through plates and metal bars to prevent vertical displacements and rotations.

FIGURE 7.29 Application of meshes on masonry samples. (Martins, A., "Seismic strengthening solutions for masonry infill walls [Soluções de Reforço Sísmico de Paredes de Alvenaria de Enchimento]," University of Minho, Civil Department, Guimarães, Portugal, 2013.)

FIGURE 7.30 Test setup. (Martins, A., "Seismic strengthening solutions for masonry infill walls [Soluções de Reforço Sísmico de Paredes de Alvenaria de Enchimento]," University of Minho, Civil Department, Guimarães, Portugal, 2013.)

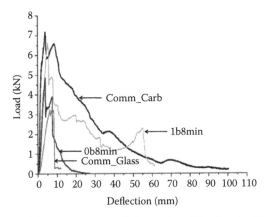

FIGURE 7.31 Load–deflection response of the meshes. (Martins, A., "Seismic strengthening solutions for masonry infill walls [Soluções de Reforço Sísmico de Paredes de Alvenaria de Enchimento]," University of Minho, Civil Department, Guimarães, Portugal, 2013.)

The typical results of the bond tests are presented in Figure 7.31. The behavior of mesh with braided rods (1b8min) presented very similar behavior in relation to the commercial mesh with carbon fibers. The maximum force was almost the same, but the postpeak behavior was very ductile.

The mesh constituted with rods without roughness presented a fragile behavior. These meshes presented transverse and longitudinal interlacing of rods, providing the locking of the mesh when it was pulled out. However, this effect is much more relevant in the case of the 1b8min mesh due to the additional roughness of each rod. The commercial solution with glass fiber presented a fragile behavior because the meshes presented total damage before they were pulled out from the mortar.

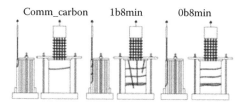

FIGURE 7.32   Failure scheme showing typical crack pattern. (Martins, A., "Seismic strengthening solutions for masonry infill walls [Soluções de Reforço Sísmico de Paredes de Alvenaria de Enchimento]," University of Minho, Civil Department, Guimarães, Portugal, 2013.)

The crack pattern of the produced meshes presented the formation of multiple cracks associated with the redistribution of forces, presenting also the fragmentation and detachment of the mortar in many areas of the plastering. For the commercial carbon mesh, a more localized failure was obtained (Figure 7.32). This difference can also be associated with the global interlocking that the produced meshes exhibited due to the interlacing of the rods, and this was not visible in the smooth commercial meshes.

7.3.5   STRENGTHENING OF BRICK MASONRY UNDER FLEXURE

The major aim of developing BCRs is the retrofitting of masonry infill walls submitted to flexural loading, to (1) increase the strength, (2) increase the deformations, and (3) control damage to avoid the abrupt failure and out-of-plane projection of masonry. Therefore, an experimental campaign was defined so that an assessment of the retrofitting effectiveness of the braided rod mesh on brick masonry under flexure could be evaluated.

Samples of brick masonry were composed of brick units mostly used in the construction of masonry infill walls in Portugal. Masonry specimens with dimensions of 300 mm × 200 mm × 150 mm (length × height × thickness) were considered. The geometry was defined according to EN-1052-2-99 (1052-2 1999) with respect to simple bending in a plane perpendicular to the bed joints. The laying of the bricks was carried out with general-purpose mortar M10, but for the finishing a rendering special mortar was used. The construction of masonry specimens was made by an experienced mason to reproduce similar workmanship used in current structures, as followed before. The samples were kept under relatively stable conditions of temperature and humidity inside the laboratory. The rendering of the specimens with the introduction of strengthening meshes was made 2 weeks after the construction of the specimens. This was done by applying a thin layer of mortar, with subsequent placement of the reinforcing mesh embedded in a new layer of rendering mortar, giving a total thickness of about 20 mm. Moreover, even in the unreinforced walls the rendering was applied with a thickness of 20 mm.

Two different types of reinforcement were produced to have a means of parametrical comparison. They are named as 2G and 4G, according to the number of yarns of fiberglass composing the core, namely, two (2G) or four (4G). The choice of fiberglass instead of carbon is due to its lower cost, better availability, and better behavior in terms of ductility. The meshes were assembled according to the defined measures, with a spacing of 60 or 30 mm. The shell material is composed of 15 simple yarns and a braid (composed of 8 yarns) of high-resilience polyester yarn.

One of the purposes of this research was to assess the capacity of improvement of the bonding between the reinforcement materials and the mortar matrix of masonry walls due to the use of braided materials. To have a reference, two different commercial solutions were chosen to be compared with the developed materials. A commercial bidirectional mesh is composed of alkali-resistant glass fiber (CM1) (Figure 7.33a). The black color is due to the alkali-resistant treatment, based on bitumen. It is composed in both directions of two yarns of glass fiber per element. In one direction they are close to each other, and in the other direction they are separated. The joints are somehow rigid, made by pressure and heat. The commercial unidirectional mesh (CM2) (Figure 7.33b) is composed in the main direction by two yarns of carbon fiber, slightly joined by a very fine glass fiber rolled around them. The yarns are slightly covered with a sand-like granular element, probably to increase the bonding properties. The transversal elements are composed of glass fibers, in a much lower density, probably to get better stability of the mesh rather than to give a really bidirectional behavior.

The experimental details of flexural tests on masonry are presented in Table 7.8. The experimental testing setup, instrumentation to measure the displacements, and loading procedure are presented in Figure 7.34. The flexural tests were carried out according to EN-1052-2-99 (1052-2 1999).

In terms of results, it is important to analyze the cracking load, cracking load displacement, maximum loading, and corresponding deflection and maximum deflection. Besides, it is important to (1) evaluate the postpeak behavior with increase or decrease of the loading capacity, (2) assess the increase in ultimate deformation, and (3) evaluate the failure mode focusing on the existence of load redistribution. The typical load–displacement curves obtained for each masonry specimen are presented in Figure 7.35.

The unreinforced masonry (URM) (Figure 7.35a) is characterized by a very brittle behavior, which is associated with the localized central crack involving the failure of the unit–mortar interface and the units (Figure 7.36a). This is the typical failure mode already pointed out by recent research studies (Vasconcelos et al. 2012). In average, the maximum flexural load, which corresponds to the cracking flexural load, presents an average value of 8.48 kN, corresponding to a flexural stress in the perpendicular direction to the bed joint, f_{xi}, of 0.45 MPa. The scatter of the flexural cracking load is low, which validates the accuracy of the experimental results.

As expected, the material is very stiff, with a very low displacement until the peak load is attained. Even if the loading rate was set to the minimum possible

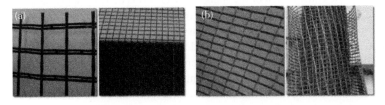

FIGURE 7.33   (a) Commercial glass mesh and (b) commercial carbon mesh. (Martins, A., "Seismic strengthening solutions for masonry infill walls [Soluções de Reforço Sísmico de Paredes de Alvenaria de Enchimento]," University of Minho, Civil Department, Guimarães, Portugal, 2013.)

TABLE 7.8
Types of Specimens Tested in Out-of-Plane Flexure

Retrofitting Material	Reinforcement Density (g/m²)	Spacing (mm)	Tensile Capacity (kN)
2G#6	57	60	9.2
2G#3	114	30	12.0
4G#6	114	60	18.4
CM1	225	25	27.4
CM2	200 (Unidirectional)	20	57.1

Source: Martins, A., "Seismic strengthening solutions for masonry infill walls [Soluções de Reforço Sísmico de Paredes de Alvenaria de Enchimento]," University of Minho, Civil Department, Guimarães, Portugal, 2013.

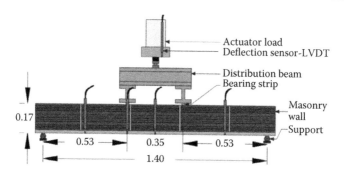

FIGURE 7.34 Experimental details and test setup (dimensions in meters). (From Martins, A., "Seismic strengthening solutions for masonry infill walls [Soluções de Reforço Sísmico de Paredes de Alvenaria de Enchimento]," University of Minho, Civil Department, Guimarães, Portugal, 2013.)

value, the samples broke in about 3 minutes, presenting an abrupt failure without any postpeak response. The responses of the specimens strengthened with BCR meshes present a similar behavior until the flexural cracking load is attained (Figure 7.35b through d), resulting in the opening of the first flexural crack. In average, the flexural cracking load obtained in the specimens 2G#6, 4G#6, and 2G#3 is similar to the load recorded in the URM specimens. This means that the developed meshes are not very effective in increasing the flexural cracking load. Besides, it is seen that after the first crack is opened there is a considerable reduction of flexural load in case of specimens strengthened with BCR meshes, whereas the reduction is more controlled in case of commercial mesh CM1 and almost absent in case of CM2 mesh. Notice that the commercial mesh CM1 is composed of the same base reinforcing material as the developed materials, that is, glass fibers. However, the reinforcement ratio is considerably higher compared to developed braided materials. On the other hand, the small spacing among the yarns guarantees a very uniform behavior. There is a very good redistribution of the load when CM2 mesh is added, with the development of several thin cracks without a definite opening of any of them. The opening of these

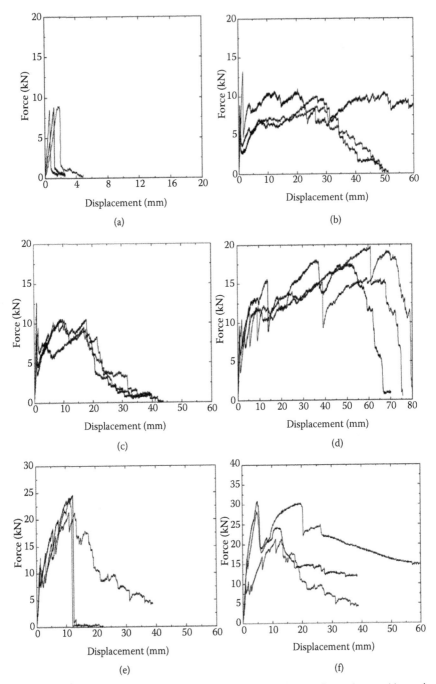

FIGURE 7.35 Typical force–displacement diagrams obtained in flexural tests: (a) unreinforced masonry, (b) masonry strengthened with mesh 2G#6, (c) masonry strengthened with mesh 4G#6, (d) masonry strengthened with mesh 2G#3, (e) masonry strengthened with mesh CM1, and (f) masonry strengthened with mesh CM2.

cracks implies reduction of the load-bearing capacity, even if an increasing tendency until the maximum load of about 23 kN is observed. The peak load is attained for a deflection of about 12 mm.

After the reduction of the flexural load, there is a recovery to a value close to the flexural cracking load in case of specimens strengthened with 2G#6 and 4G#6. In case of mesh 2G#3, there is a progressive increase in the flexural load to more than double the flexural cracking load. The great advantage of BCR meshes (2G#6 and 4G#6) is the improvement in behavior after the cracking load, as the specimens are able to recover a great percentage of the flexural resistance for high values of lateral deformation, even if no advantage was observed in the crack distribution (Figure 7.36a and b) in relation to URM. On the other hand, the mesh 2G#3 results in effective force redistribution, resulting in more distributed crack patterns (Figure 7.36c).

It is observed that the BCR meshes present a more ductile behavior than the mesh CM1, in which two specimens failed abruptly after the achievement of the maximum load. In case of the specimens reinforced with the BCR mesh 2G#3, the increase in flexural load is accompanied by the increase in deformation. Contrary to the other BCR meshes, in this case it is possible to observe that there is a more effective redistribution of forces as the mesh is also effective in increasing the resistance (Figure 7.36d). It should be noticed that the ductility measured as a function of the maximum displacement is also higher in case of BCR mesh 2G#3, being practically similar to the ultimate deformation exhibited by the specimens strengthened with mesh CM2. The use of this commercial solution leads to an enhancement of the first cracking load of around 20% compared to the unreinforced samples. The most remarkable feature of its behavior is the quick redistribution of the loads and

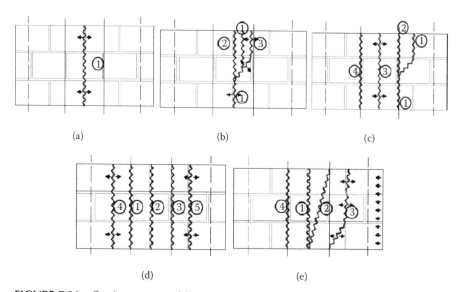

(a)   (b)   (c)

(d)   (e)

FIGURE 7.36 Crack patterns at failure (the numbers indicate the sequence of opening of the cracks): (a) masonry strengthened with mesh 2G#6, (b) masonry strengthened with mesh 4G#6, (c) masonry strengthened with mesh 2G#3, (d) masonry strengthened with mesh CM1, and (e) masonry strengthened with mesh CM2.

the development of several thin cracks. The formation of cracks does not imply the dropping of the load-bearing capacity but just a lowering of the stiffness (cracked stiffness). In this case, a distributed crack pattern is observed, even if the failure sliding of the carbon fibers is recorded.

7.4　FUTURE TRENDS

Shaping the cities for the challenges of the future is indeed a major focus of both the scientific community and the business world. The use of new materials plays an important role in the creation of cities of the future, especially in facing the challenge of urban population growth. The scientific community has been putting in much effort to develop new materials that can keep up with the evolution of human society. In the past decades, four major innovative materials have been developed, namely, self-healing, shape memory, self-monitoring, and high strength to weight ratio materials.

Nanotechnology presents material properties and characteristics that are not available at the macroscale, for instance, the capability to create materials with ultra-high mechanical properties combined with low weight. In the future, these kinds of materials will be seen in the braided structures for several construction applications.

Electronics is, nowadays, applied in almost all aspects of human life. New materials for electronics, especially fiber-based ones, are being studied and developed to overcome the problems of traditional ones. Fibers present several advantages in this area due to their flexibility and easy integration with several materials. Carbon, on its allotropic conductive phase, is being produced at both nanoscale and microscale to achieve better electrical properties. These kinds of novel materials will be explored in braided structures and composites to give monitoring capacity to all constructions, improving their safety and durability.

The increase of scientific activity in the field of shape memory materials leads to the development of flexible shape memory polymers and fibers. In the future, these shape memory materials will be used in braiding technology to give this outstanding behavior to fiber-based structures and composites, and concomitantly to their final application.

Last but not least, one of the biggest innovations in materials science is the invention of regenerative capacity of materials, leading to the development of so-called self-healing materials. These materials will play an important role in the future in constructions, providing them the ability to self-repair after being damaged by natural or other causes, and braided structures and composites can prove to be an important means for introducing this property to civil structures.

REFERENCES

1052-2, BS EN. "Methods of test of masonry—part 2: determination of flexural strength." 1999.

Al-Chaar, G., M. Issa, and S. Sweeney. "Behavior of masonry-infilled nonductile reinforced concrete frames." *Journal of Structural Engineering* 128(8) (2002): 1055–1063.

ACI Committee 440. *State-of-the-Art Report on Fiber Reinforced Plastic (FRP) Reinforcement for Concrete Structures - ACI 440R-96.* American Concrete Institute, 2002.

Balaguru, P., A. Nanni, and J. Giancaspro. *FRP Composites for Reinforced and Prestresses Concrete Structures.* New York: Taylor & Francis, 2008.

Balendran, R. V., T. M. Rana, T. Maqsood, and W.C. Tang. "Aplication of FRP bars as reinforcement in civil engineering structures." *Structural Survey* 20 (2002): 62–72.

Bank, L. C. *Composites for Construction—Structural Design with FRP Materials.* New Jersey: John Wiley & Sons, Inc., 2006.

Benmoktane, B., O. Chaallalt, and R. Masmoudi. "Glass fibre reinforced plastic (GFRP) rebars for concrete structures." *Construction and Building Materials* 9(6) (1995): 353–364.

Bertolini, L., B. Elsener, P. Pedeferri, and R. Polder. *Corrosion of Steel in Concrete: Prevention, Diagnosis, Repair.* Weinheim, Germany: John Wiley & Sons, 2004.

Ceroni, F., E. C. M. Gaetano, and M. Pecce. "Durability issues of FRP rebars in reinforced concrete members." *Cement & Concrete Composites* 28 (2006): 857–868.

Cui, Y., M. S. M. Cheung, B. Noruziaan, S. Lee, and J. Tao. "Development of ductile composite reinforcement bars for concrete structures." *Materials and Structures* (RILEM) 41 (2008): 1509–1518.

Cunha, F. Development of braided composite rods to be used in the strengthening of masonry walls [*Desenvolvimento de uma estrutura com materiais fibrosos para ser utilizada como reforço de paredes de alvenaria*]. Guimarães, Portugal: University of Minho, Textile Department, 2012.

Micelli, F., and A. Nanni. "Durability of FRP rods for concrete structures." *Construction and Building Materials* 18 (2004): 491–503.

Fangueiro, R. *Fibrous and Composite Materials for Civil Engineering Applications.* Cambridge, United Kingdom: Woodhead Publishing, Ltd., 2011.

Fangueiro, R., and C. Gonilho-Pereira. "Fibrous materials reinforced composite for internal reinforcement of concrete structures." In *Fibrous and Composite Materials for Civil Engineering Applications*, 216–249. Cambridge, United Kingdom: Woodhead Publishing, Ltd., 2011.

Fangueiro, R., Rana, S., and Correia, A.G. "Braided Composite Rods: Innovative Fibrous Materials for Geotechnical Applications." *Geomechanics and Engineering* 5(2013): 87–97.

Federation International de Beton. *Fib Bulletin 40—FRP Reinforcement in RC Structures.* Technical Report, Federation Internationale du Béton Task Group 9,3. 2007.

Gómez, J.M. *Innovative Retrotting Materials for Brick Masonry Infill Walls.* Guimarães, Portugal: University of Minho, Department of Civil., 2012.

Hampton, F., F. K. Ko, and H.G. Harris. "Creep, Stress, Rupture, and Behavior of a Ductile Hybrid FRP for Concrete Structures." 12th International Conference on Composite Materials. Paris, France, 1999. p. 731.

Hampton, F. P., and Frank K. Ko. "Earthquake Resistance of Concrete Columns Reinforced with Ductile Hybrid Fiber Reinforced Polymer (DHFRP)." 16th International Conference on Composite Materials. Kyoto, Japan, 2007.

Harris, G. H., W. Somboonsong, and F. K. Ko. "New ductile hybrid FRP reinforcing bar for concrete structures." *Journal of Composites for Construction* 2 (1998): 28–37.

Harris, H. G., F. P. Hampton, S. Martin, and F. K. Frank. "Cyclic Behavior of a Second Generation Ductile Hybrid Fiber Reinforced Polymer (D-H-FRP) for Earthquake Resistant Concrete Structures." 12th World Conference on Earthquake Engineering (12WCEE). Auckland, Australia, 2000.

Hota, V. S. GangaRao, Narendra Taly, and P. V. Vijay. *Reinforced Concrete Design with FRP Composites.* Broken Sound Parkway NW: CRC Press, 2007.

International Bureau for the Standardisation of Man-Made Fibers. *Terminology of Man-Made Fibers.* Brussels, Belgium: BISFA, 2006.

Karbhar, I. V. M. "Introduction: the use of composites in civil structural applications." In *Durability of Composites for Civil Structural Applications*, by Vistasp M. Karbhari. Abington, United Kingdom: Woodhead Publishing, Ltd., 2007.

Karbhari, V. M. *Use of Composite Materials in Civil Infrastructure in Japan.* Maryland: International Technology Research Institute—World Technology (WTEC) Division, 1998.

Katz, A. "Bond to concrete of FRP rebars after cyclic loading." *Journal of Composites for Construction* 4(3) (2000): 137–144.

Katz, A. "Bond mechanism of FRP rebars to concrete." *Materials and Structures* 32 (1999): 761–768.

Lam, H., F. Hampton, Frank K. Ko, and H. G. Harris. "Design Methodology of a Ductile Hybrid Kevlar-Carbon Reinforced Plastic for Concrete Structures by the Braidtrusion Process." 13th International Conference on Composite Materials. Beijing, China, 2001.

Lourenço, P.B., G. Vasconcelos, P. Medeiros, and J. Gouveia. "Vertically perforated clay brick masonry for loadbearing and non-loadbearing masonry walls." *Construction and Building Materials* 24(11) (2010): 2317–2330.

Marques, A. T. "Fibrous materials reinforced composites production techniques." In *Fibrous and Composite Materials for Civil Engineering Applications*, by R. Fangueiro. Cambridge, United Kingdom: Woodhead Publishing, Ltd., 2011.

Marques, P. *Design of Fibrous Structures for Civil Engineering Applications* (master's thesis). Guimarães, Portugal: University of Minho, 2009.

Martins, A. Seismic strengthening solutions for masonry infill walls *Soluções de reforço sísmico de paredes de alvenaria de enchimento*. Master thesis, Guimarães, Portugal: University of Minho, Civil Engineering Department, 2013.

Moreno, M., W. Morris, M. G. Duffó, and G. S. Alvarez. "Corrosion of reinforcing steel in simulated concrete pore solutionsin simulated concrete pore solutions." *Corrosion Science* 46 (2004): 2681–2699.

Nanni, A., T. Okamoto, M. Tanigaki, and S. Osakada. "Tensile properties of braided FRP rods for concrete reinforcement." *Cement & Concrete Composites* 15 (1993): 121–129.

Pastore, C. M., and F. K. Ko. *Braided Hybrid Composites for Bridge Repair*. National Textile Center Annual Report, Philadelphia and Drexel University, 1999.

Pultrall, Inc. *Pultrall*. Publi-Web.net. http://www.pultrall.com/vrod.asp?languageid = 2 (accessed January 5, 2013).

Rana, S., Zdraveva, E., Pereira, C., Fangueiro, R., and Correia, A. G. "Development of Hybrid Braided Composite Rods for Reinforcement and Health Monitoring of Structures", *The Scientific World Journal* 2014 (2014): 1–9.

Reis, V. L. F. Construction of reinforced concrete structures with FRP rods, Master thesis, Faculty of Engineering, University of Porto, 2009.

Rosado Mérida, K. P., Rana, S., Gonilho-Pereira, C., and Fangueiro, R. "Self-Sensing Hybrid Composite Rod with Braided Reinforcement for Structural Health Monitoring", *Materials Science Forum* 730–732 (2012): 379–384.

Somboonsong, W., F. K. Ko, and H. G. Harris. "Ductile hybrid fiber reinforced plastic reinforcing bar for concrete structures: design methodology." *ACI Materials Journal*, 95(6), 655–666, 1998.

Vasconcelos, G., F. Cunha, R. Fangueiro, and S. Abreu. "Retrofitting Masonry Infill Walls with Textile Reinforced Mortar." 15th World Conference on Earthquake Engineering. Lisbon, Portugal, September 24–28, 2012.

Waldron, P. "The use of FRP as embedded reinforcement in concrete." *FRP Composites in Civil Engineering—CiCE 2004*, Adelaide, Australia, 2004.

Wang, Y. C., P. M. H. Wong, and V. Kodur. "An experimental study of the mechanical properties of fibre reinforced polymer (FRP) and steel reinforcing bars at elevated temperatures." *Composite Structures* 80 (2007): 131–140.

Wu, Z. "Structural strengthening and integrity with hybrid FRP composites." *FRP Composites in Civil Engineering—CICE 2004*: Seracino (Ed.), 2005.

You, Y. -J., Y. -H. Park, H. -Y., Park, and J. -S. Kim. "Hybrid effect on tensile properties of FRP rods with various material compositions." *Composite Structures* 80 (2007): 117–122.

8 Applications of Braided Structures in Aerospace Engineering

Claudio Scarponi

CONTENTS

ABSTRACT

This chapter describes the typical applications of braided composite materials in the aerospace field, such as part of airplane fuselages, stator vanes, and rotor blades (mostly for helicopters). To explain the advantages of such a new technology, this chapter also describes materials, process tools, cost-saving concepts, basic concepts of design, and a list of main structural analysis computer codes used for this purpose.

8.1   INTRODUCTION

For several decades, traditional composite laminates, made by laying up diverse angle-oriented plies, have been manufactured for use as engineering materials. Enterprises and factories have acquired good skills and know-how for producing multilayered composites with cutting-edge techniques and machinery.

However, these types of materials show some weaknesses and deficiencies. Due to insufficient bonding between the various plies of which they are made of, classic laminates exhibit poor strength against out-of-plane interlaminar stresses with consequently low toughness and damage tolerance, which are fundamental aspects for aerospace applications.

For instance, a small incident due to delamination happened to a Boeing 757 (flight 1990 of American Airlines which was flying from Miami to Orlando in July 2010). Delamination affected the slat of the right wing, in proximity of the leading edge. Slats are usually made using "sandwich panels" made of graphite/epoxy layers bonded to a Nomex honeycomb core. In Figure 8.1, it is possible to see the damage in two different phases, in flight and after landing, with evidence of the damage propagation.

Moreover, complex structures often cannot be shaped as final parts due to manufacturing constraints; they need to be joined together with secondary operations, such as bonding by the use of structural adhesives or mechanical fastening by the use of rivets or bolts. Of course, the use of fasteners implies drilling a large number of holes, which induce weakness points and stress concentrations in structures to be accounted for, with evident constraints of additional time, weight, and costs.

Hence, over the years a through-the-thickness (TTT) reinforcement (Figure 8.2 shows a stitching machine) has been developed to enhance the performance of the part in presence of interlaminar stresses (three components of the stress tensor), known as two shear stresses between the plies (ISS along x–z and y–z axes, following the usual nomenclature in the scientific literature) and one normal stress along the so-called z direction (out-of-plane NIS). Such stresses are critical for laminated composites, producing delamination onset and propagation.

The so-called "stitching" technique consists of adopting z-directional extra reinforcements, to bond the plies together. Lockstitch, chain stitch, tufting, and z-pinning are only some of the typical stitching techniques [1]. The first three techniques utilize needle machines and threads (at the beginning using cotton, nowadays using glass or Kevlar), whereas z-pins are made by carbon fibers. In case of delamination, z-axis threads and z-pins operate constraining its propagation, bridging the layers between them.

FIGURE 8.1   Slat delamination.

FIGURE 8.2   National Aeronautics and Space Administration advanced stitching machine at the Marvin B. Dow Stitched Composite Center for the production of Boeing's low-cost composite wing structures.

It is obvious that needle pass, pitch line, fiber tension, and other fundamental parameters play an important role. Incorrect setting of any of these parameters can result in degradation and lower performance of the composite structure, in comparison with those produced using the traditional layup techniques, vacuum bags, and autoclaves. For example, excessive tension of the threads could produce resin pockets, unacceptable friction between in-plane fibers, and local microdamages, capable of resulting in flaws and cracks. Considering the presence of threads also in the thickness direction, such a technique can be considered as a precursor of braiding technology.

Braided structures are considered to be the natural development of stitched structures not only for the improvement of mechanical characteristics, such as higher toughness, better impact strength, and better stopping of crack propagation, but also for two more fundamental reasons, cost saving and production flexibility.

The high level of automation, high reproducibility, and capacity to produce complex preforms very near to the net final shape with good standardization levels, strongly reduce the labor cost in comparison to the traditional processes. The complex, giant machinery permits a high level of flexibility in the sense that different fibers can be used, also for hybrid production, with different fiber diameters, fiber angles and areal weights. More details of braiding process are given in Section 8.4.

8.2   TYPES OF BRAIDED STRUCTURES

8.2.1   TWO-DIMENSIONAL STRUCTURES

Initially, braiding was performed using a two-dimensional (2D) technique. The 2D braiding process to manufacture rope like structures, such as shoelaces, has traditionally been used for years. Control of a traditional 2D braider is a relatively simple task [2]. This includes two sets of yarn carriers rotating on a circular track, one rotating in the clockwise direction and the other in the counterclockwise direction (Figure 8.3a and b and Figure 8.4); during this process, yarns interlace with one another to form a tubular braided structure.

The movement of the carriers (large machines have hundreds of them) is accomplished through the use of horn gears (Figure 8.5), which allow the transfer of yarns from one gear to another.

Classical 2D braiding consists of the formation of structures by interweaving a set of yarns symmetrically about an axis. Products can be flat, tubular, rectangular, and so on, woven around a mandrel of any shape. So, braiding can also be used to produce complex shapes, depending on the mandrel around which the yarns interlace over.

Mandrels can be cylindrical, tapered, or slightly curved. Although this could be considered as three-dimensional (3D) textile forming, in the sense of a surface in 3D space, it is called 2D braiding nonetheless because of the lack of TTT reinforcement.

It is noteworthy that 2D braiding over a mandrel is a very consolidated process; moreover, it is inexpensive, is flexible, and can be run at production speeds to get low-cost, high-fiber-volume, and high-toughness composites.

Specialized braiding machines [3] can produce biaxial and triaxial braids (Figures 8.6 and 8.7). The first type usually consists of angle ply fabrics with particular patterns [4], known as diamond, regular, and Hercules (Figure 8.8), and plain weave or satin as fabric styles.

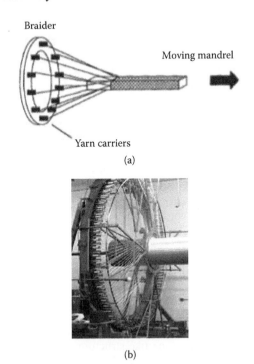

(a)

(b)

FIGURE 8.3 Disposal of fibers in a braiding machine: (a) schematic diagram and (b) in a working machine.

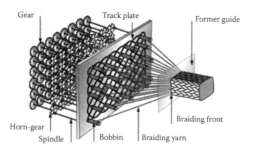

FIGURE 8.4 Conceptual drawing of a braiding machine.

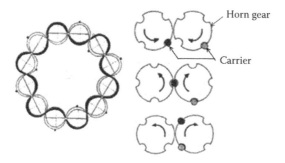

FIGURE 8.5 Horn gears and yarn carriers.

FIGURE 8.6 Biaxial two-dimensional braids.

FIGURE 8.7 Two-dimensional triaxial braids.

FIGURE 8.8 Type of braid styles.

Triaxial braids contain 0° tows along the preform length, so that in-plane axial properties are enhanced. The braid angle, or bias angle along the preform axial direction, ranges from 10° to 80° depending on the desired properties. Of course, the choice of style strongly depends on the required drapability of the material. Drapability means the quality of the material to be laid into a mold of complex shape.

Depending on the required characteristics, diamond is excellent for its structural stability; but the other ones are recommended for their reduced crimps and wrinkles, due to a more limited intertwining of the tows.

All these diverse kinds of manufacturing techniques and classes of obtainable configurations, along with the tailoring of the mandrel to the desired final shape, bring an enormous variety of feasible structures and shapes designed for specific requirements. For example, the fan case would prefer a configuration with high braid angles because of its requirement of containing and absorbing unpredictable blade failures, instead of withstanding an outstanding axial load.

8.2.2 THREE-DIMENSIONAL STRUCTURES

The next technology, which is is still performing, is to fabricate 3D fabrics and near to net shape form of reinforcements. So, nowadays the most advanced works are carried out on 3D preforms of reinforcement; 3D braiding and 3D weaving (Figure 8.9) are the most important kind of 3D composite forming techniques [2].

A preform is a preshaped fibrous reinforcement incorporating various structural details and formed to the desired shape before being placed in moulds to be filled with resin (of course, in case of absence of prepreg threads). The two main processes adopted for these kinds of materials are vacuum bag and autoclave (for prepreg threads) and resin transfer molding (RTM) (Figures 8.10 and 8.11) or resin film infusion (RFI) (Figure 8.12), both applicable to dry preforms.

The 3D weaving introduces further out-of-plane reinforcements, interlaced with each other and with the traditional warp and weft yarns in the longitudinal and transverse directions. An advantage of this process is that it can be performed on conventional looms and does not require significant costs to develop specialized machinery. However, its main limitation is the possibility of producing mostly 3D flat fabrics. So, the appealing characteristic of forming net-shaped fibrous parts is not consistent.

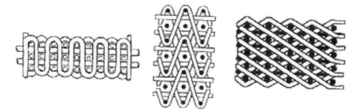

FIGURE 8.9 Three-dimensional (3D) braids.

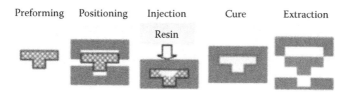

FIGURE 8.10 Scheme of resin transfer molding process.

FIGURE 8.11 Resin transfer molding process: example of part production. Fuselage panel.

Nowadays, 3D braided composite materials and their different complex machines are being developed increasingly and will represent the state of the art for composite production in a few years.

Summarizing, the advantages of 3D braided preforms are as follows:

- Superior toughness and fatigue resistance respect to filament wound composites (intertwining points between tows act as cracks and delamination arresters)
- Net shape preform
- Fiber continuity
- Reduction of cuts and mechanical joints
- Automation and costs reduction

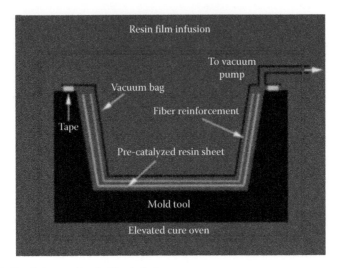

FIGURE 8.12 Scheme of resin film infusion process.

FIGURE 8.13 Three-dimensional braiding machine horn gear.

- Continuous selvedge of flat fabrics and seamless shape of tubular forms
- Fabrication over a wide range of angles of interlacement along the reference plane and the z axis

Some disadvantages are as follows:

- The production of small-scale preforms relative to the size of the machinery.
- Resupply and restocking of yarns are necessary, so the production of long preforms is slower.

The 3D braiding machines, able to produce 3D preforms, are classified into two main classes: horn gear types [2] and "track and column" types; only the first one is described for simplicity (Figure 8.13). More detailed discussions on braided structures and machines will be found in Chapter 1 of this book.

This kind of machine uses a large number of traditional horn gears for carriers' propulsion, by arranging them in a square or in a cylindrical frame. This allows the yarn carriers to transfer between adjacent tracks, thus forming a multilayered fabric with yarns interlocking closed layers.

8.3 TYPES OF APPLICATIONS

8.3.1 ORIGIN

The origin of braiding process is placed by many encyclopedias at thousands of years ago. Ancient Chinese and Japanese documents record that braiding was in use even before 4000 BC. The first braiding machine patent was issued in 1748 in Manchester, England. Since then, this kind of machines has been used primarily in the fabrication of cordage and ropes. In the late 1960s of the twentieth century, braiding began to be used in the fabrication of composite materials. The first components produced were rocket nozzles (Figure 8.14) to replace high-temperature metallic alloys and save up to 30%–50% in weight.

At the time, only few motor components were made, although it demonstrated the capability of the braiding process to produce intricately shaped components. Immediately afterward, other 3D composite–forming processes started developing, such as weaving and knitting.

In the mid-1980s, the National Aeronautics and Space Administration (NASA) Advanced Composite Technology (ACT) program [2,5] impelled the development of 3D fibre reinforced plastics composites. From 1985 to 1997, the ACT program funded contracts with major aerospace companies for textile structural components and funded a concurrent research effort by government, industry, and universities to provide an engineering and science base for textile composites. The scope was to give impetus to a more rapid transition of such a technology into production of aircraft parts (Table 8.1).

ACT also helped in improving and developing the mechanic theory and modeling of 3D composites. In particular, ACT focused its attention on primary structures of commercial transport aircrafts. Some of the participants and contractors of ACT were Boeing, McDonnel Douglas Corp., Lockheed Martin Corp., Northrop Grumman Corp., Atlantic Research Corp., NASA Langley Center, Drexel University, University of Delaware, and so on.

TABLE 8.1

Applications of 2D and 3D Braided Structures in the Aerospace Field

Braided Structures	Use in Aerospace Structures
2D braid	Propellers, rotors, helicopter blades, wind turbines blades
3D sock braid	Propellers, blades
3D braid	Wing stringers, stator vanes, fuselage, wings

FIGURE 8.14   Braided rocket nozzle.

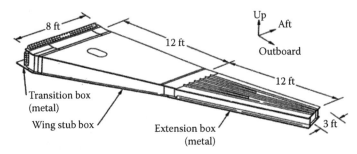

FIGURE 8.15   Stitched wing stub box.

The most important result of ACT was a completed development of a stitched/RFI stub box (4 m main wing tapered spar) by McDonnel Douglas, which was revealed as being cost-competitive and even easily repairable (Figure 8.15); it is not a surprise that the first important application in aeronautics was made by the braided preform.

8.3.2 Fan Blade Containment Case

Since the 1990s to the beginning of the twenty-first century, NASA Glenn Center and A&P technology (leader in braiding technology) started to research and develop a new type of weight-saving and impact-tolerant fan blade containment using braiding.

They have demonstrated its commercial feasibility and structural capability, after various tests and a new manufacturing procedure, using vacuum assisted resin transfer molding (VARTM). This containment case (Figure 8.16) is fundamental in case of a failure and a fan blade out event, because it must contain the failed blade and the consequent possible impacts and crack propagation [6,7]. In fact, one of the many challenges faced by the jet engine designers is to contain a failed fan blade within the engine so that it threatens neither passengers nor airframe.

The fan case structure has to not only withstand the blade impact but also retain its structural integrity while the engine is shutting down. After an engine loses a blade, the loads on the fan case rise well above those experienced in normal flight, due to fan imbalance. During engine shut down, in typically 15 seconds, cracks can

FIGURE 8.16 Fan blade containment case.

propagate rapidly in the fan case from the damage caused by the impact of the broken blade. If the fan case fails completely, then the consequences for the engine and aircraft could be catastrophic.

There are two different approaches to blade containment currently in common use. One, the soft-wall fan case, features a casing of aluminum overwound with dry aramid fibers. It is designed to permit a broken blade to pass through its aluminum component, where it is stopped and contained within the external aramid fiber wrap.

The second type, the hard-wall fan case, is aluminum only and is designed to reflect the blade back into the engine. The hard-wall concept enables designers to improve engine aerodynamics by building a case with a smaller radial envelope, as no "dead" space is required for deflection of an aramid wrap. However, a serious risk is that the broken blade may pass through the engine, with the possibility for secondary damage well in excess of that experienced in the soft-wall design.

The key material properties are the ductility of the metal cases and, in the case of the soft-wall system, the energy-absorbing capabilities of the aramid fibers.

The solution of using a triaxial braid permitted the production of a lighter structure, with the necessary characteristics of strength and ductility.

The key component of this new system is its reinforcement, developed by A&P technology. The braid architecture is very resistant to delamination between plies. Because all fiber orientations are present within each ply, all plies are identical. The stiffness match at ply interfaces, therefore, results in inherent structural toughness. Furthermore, the triaxial fiber architecture limits linear crack propagation, confining failure to very limited regions.

8.3.3 Aircraft Propellers

The reasons behind the choice of braided materials even in propellers are the same as those listed earlier for the fan case.

The propeller is one of the most critical parts of an aircraft engine, as it is highly exposed to external environment and plausible violent impacts due to foreign object damage (FOD), such as bird strikes. Figure 8.17 represents an airplane propeller, protected by the U.S. Patent n. US8221084 B2, year 2008. The particular n. 32 represents the braided sock used for the construction of the central body.

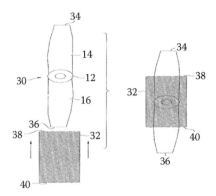

FIGURE 8.17 Aircraft propeller, draft.

8.3.4 STATOR VANES

The stator vane (Figure 8.18) is an important part of a turbojet engine. The fluid must flow in stationary conditions and must be correctly directed to adjacent rotating fan blades, so geometry must be strictly controlled. So, precise airfoil shape is very crucial. Other requirements for this application are high fatigue strength and impact strength, to avoid the possible ingestion of the so-called FOD.

Honeywell performed a trade-off between various forms of carbon fiber reinforcement available for stator construction and discovered braided reinforcement developed by A&P Technology. Vanes molded using A&P Technology's braided carbon fiber and RTM process could perfectly respond to the strong requirements of vanes, dramatically reducing part cost with respect to unidirectional carbon prepreg and autoclave molding process. Best of all, they could be perfectly tailored to the shape of the mold.

The preform has been obtained by braiding an aramid sleeve and then overbraiding the aramid sleeve with a carbon fiber sleeve. Both braids are computer controlled, to allow thickness variations in the airfoil shapes. The resulting sleeve, thick in its central portion and thin at the edges, is trimmed to the desired length. Then, the molder places the preforms in a multicavity RTM tool (mold). Once the mold is closed, epoxy resin is injected and cured. The resulting vane is ready for use with only minor trim and deflashing operations.

A&P Technology's predictive software, sophisticated braiding techniques, and mandrel designs allow reinforcements to be tailored to the precise dimensions and structural requirements of the product. Honeywell stator vanes made with A&P Technology's braided preforms are now in production and flying in engines all over the world.

8.3.5 OTHERS

Hence, over recent years braiding has been applied in secondary structures though complex and specialized parts where damage tolerance, impact resistance, energy absorption, and high degree of detailed and intricate preform shapes are essential. A typical application in this field is the window frame, which is shown in Figure 8.19.

FIGURE 8.18   Stator vane.

FIGURE 8.19   Window frames.

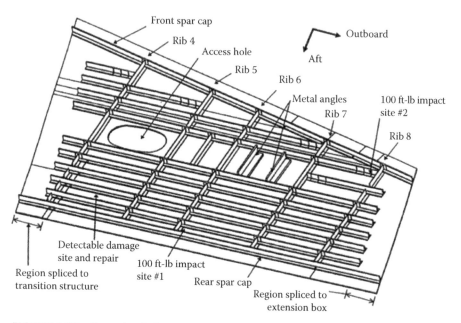

FIGURE 8.20   Example of wing structure.

Even control surfaces and wing structures (Figure 8.20), already produced industrially using composite materials, are now under study; Bombardier, the world's third civil airframe manufacturer, developed a complete composite wing flap in collaboration with A&P Technology, which provided the braided carbon fiber sleeves, placed over an internal flap-shaped mandrel [8].

As a typical transfer of this kind of technology, parts for Formula One racing cars and the entire frame of the advanced models of Mercedes, Audi, and Volkswagen are fabricated by braided materials, thanks to the excellent crash energy-absorbing characteristics.

Perhaps in the future, we will use this technology for the substitution of traditional carbon fiber tennis rackets, skies, fishing pools, sail boat masts, and so on.

8.4 REQUIREMENTS, FIBERS, AND PROCESSES

Both Boeing and Airbus, the two giant companies of aeronautical construction, are going ahead with the use of braided composites in the new generation of airplanes B 787 and A350/A380, respectively. The excellent performance (lower fuel consumption) and the reduction of labor cost are the main reasons behind this.

From the structural point of view, in addition to the well-known characteristics of traditional composite structures, braided composites fulfill the essential requirements of aerospace applications as follows:

- As a structure is exposed to high fatigue cycles, cracks will propagate through the matrix of filament-wound or unidirectional prepreg laid-up structures. Whereas when microcracking will occur in a braided structure, the propagation is arrested at the intersections of the reinforcing yarns. That is why braid is the reinforcement choice for aircraft propellers and stator vanes in jet engines.
- Braid also greatly improves interlaminar shear properties when nested together with other braids. While interlaminar adhesion is not different from other reinforcement products, the layers move together. As a result, it is very rare for cracks to form and propagate between layers of braided reinforcement. Because braids are produced using the bias fibre directions, they provide very efficient reinforcement for parts that are subjected to torsional loads. So, braid is an ideal reinforcement for drive shafts and other torque transfer components, such as flanged hubs.
- Excellent behavior to impact loads and excellent toughness is due to the high deformation capacity and no-fragmentation quality.
- Flexibility in the choice of fibers (glass, aramid, carbon, ceramics, and natural fibers); hybrid production; availability of different fiber diameters; and availability of a virtually infinite variety of diameters or widths, fiber angles, and areal weights.

From the production point of view, the braiding technology allows the production of complex preforms very near to the net final shape, including stringers and ribs (Figure 8.16), using very complex and giant braiding machines.

Moreover, it is possible to get higher level of automation, higher reproducibility, better standardization levels, and lower labor cost using braided composites in comparison with traditional processes.

Some other very important characteristics of braiding technology are geometrical flexibility and drapability. Braids can be easily and repeatedly adjusted to fit over molding tools or cores, accommodating straight, uniform cross-sectional forms as well as nonlinear, irregular cross-sectional components, much like a sock conforms easily to a foot. This ability to slip onto tools and cores with speed, ease, and a high degree of repeatability makes braids the ideal solution for products with changing geometries.

The most used fibers in aeronautics are carbon fibers, both high modulus and high strength. In fact, for applications in primary structures (wings, fuselages, etc.), deformation represents a more restricting requirement than strength.

Also, aramid fibers, among which DuPont "Kevlar" is the most popular, are utilized in this field, mostly in the production of multilayer plane systems, a sort of 2D braided products. It is also not unusual to use hybrid carbon/Kevlar fabrics. In Table 8.2, the main characteristics of the major fibers used for braided materials are reported.

Complex dry preforms are usually utilized for producing braided composites, thus the impregnation of fibers with resin must be performed very carefully.

One of the main processes utilized for the purpose is the aforementioned RTM, consisting, in simple words, of the application of stabilized dry preforms inside the mold and then adjusting a countermold and sending the pressurized thermoset polymer to impregnate the fibers. Of course, it is necessary to calculate the volume of resin to be injected inside the molds and to recover the excess amount of resin, usually sent to the preforms.

A variation of the process, called VARTM, uses a vacuum bag instead of the countermold, to save money, but renounces to a smooth surface on the bag side. In both cases, the polymerization can be obtained by hot molds.

Several factors have an influence on the process:

- Geometry
- Quality of molds
- Number and position of the resin gates

TABLE 8.2

Mechanical Characteristics of Fibers and Metals	Density (10^3 kg/m^3)	Modulus (GPa)	Tensile Strength (MPa)	Specific Stiffness (GPa)	Specific Strength (MPa)
E-glass	2.5	70	1700	28	680
Carbon	1.8	230 to 820	2000 to 7000	128 to 455	1111 to 3900
Aramid	1.4	130	3000	98	2140
HT steel	7.8	210	750	27	96
Aluminum	2.7	75	260	28	96

- Resin viscosity
- Dry-phase permeability
- Injection pressure
- Curing cycle

A second important process, called RFI, consists of deposition of resin over the fibers in the form of thin layers. So, the preforms are made by calculated series of dry materials and resin layers.

The polymerization is done inside the autoclave, so pressure gives good compaction and heat produces melting of resins and impregnation of fibers.

8.5   DESIGN AND DEVELOPMENT PROCESS

The design of an aerospace structure is a typical team job, coordinated by a program manager who coordinates tasks requiring different skills. He has full responsibility of the success of the design, so he has to check schedule and resources (human and economic) of the entire project, from the beginning of the concept to the shipping of the first item.

A design usually starts with a concept, followed by a preliminary phase where a trade-off among different solutions is performed so that preliminary calculations and tests to explore the possible solutions can be performed. Finally, after several audits he decides the final configuration. Then, design and fabrication processes are frozen and the go-ahead phase can start.

The dream of any manager is that schedule and activities proceed without inconvenience, but this is just a dream. In fact, it is not unusual to encounter inconveniences, obstacles, and problems concerning tools, calculations, personnel, funds, and so on, so continuous adjustments are necessary.

First of all, a wide description of all the activities must be prepared to have the necessary resources to perform the activities and to evaluate the time necessary to complete them.

It is his own job to prepare the flowchart of the activities (the so-called PERT diagram) and to identify the critical path. This is a fundamental item, because any delay of the activities inside the critical path will determine a delay of the whole program. The method is based on statistics tools, which are able to determine the timing of each activity of the project. Each phase has been discussed in more detail below.

As mentioned earlier, the first approach is the concept. Usually, the input comes from the marketing area, after consultation with customers and subcontractors. Of course, marketing and technical requirements have to be joined together, so that companies will be able to conceive the first idea of what will be the project.

At the end of this first phase, engineers know their targets and each team has its own task to accomplish. It is not rare for two or more teams to work on the same task separately to take up different approaches to the same problem; this is done especially for critical parts, like wings, carriage, rudders, and so on. When this phase ends, a joint team is made up by specialized engineers from each team of the previous phase.

Depending on the structural requirements and specifications, the team is more or less composed of the following:

- Structural engineers for statics, dynamics, fatigue analysis, thermal analysis, and so on
- Materials engineers
- Aerodynamics engineers (if necessary)
- Technology engineers, for the definition of molds, tools, materials, and so on
- Fabrication process engineers, responsible for fabrication of items and fabrication sequence
- Production engineers, in production line, responsible for time and methods
- Testing engineers (destructive and nondestructive tests)
- Quality assurance personnel
- Personnel responsible for shipping to customers

The concept becomes the project. The system is completely defined in every single section, and it is ready to be designed.

The designing phase is fundamental; design engineers, using data from aerodynamics, loads, technology, and process fabrication teams, draw a first sketch of the system.

All the codes, materials, tests, tools, instruments, and quality assurance procedures must follow international standards. Nowadays a new approach consisting risk analysis is followed, not in operations, but just at the beginning of the designing process.

Audits with the authority are usual along the designing process to respect the certification procedures.

Structural design is performed by the use of specialized codes. The so-called finite elements technique is universally used for aerodynamics, engines, structures, and so on. The most important codes for structural analysis are MSC NASTRAN, CATIA, ANSYS, ABACUS, LN-DYNA, and so on.

Here, finite element design methodology has been discussed in simple words. The basic concept is to divide the structure into very small elementary parts, depending on geometry and structural behavior. So, we have 3D or 2D or one-dimensional elements described (for instance) by cubes, plates, and beams, respectively. Each element is determined by points, called nodes (almost eight for 3D elements, three for plane elements, and two for linear elements). Usually, 3D elements are cubes, excepting boundary and corner elements, which usually are parallelepipeds with a triangular base.

The discretization of the structure in nodes produces a mesh. The engineer determines the typology of elements constituting the structure and defines the mesh with the help of specialized preprocessors (PATRAN, for instance), which are able to optimize the position identification of the nodes. Each node has a number, and each structural element (defined by a connection of a set of nodes) also has a number. The model also takes into account the angles of the fibers in a 3D space, so the help of specialized preprocessors is very important.

Input data depend, of course, on the requested analysis. For instance, in the case of structural robustness only mechanical characteristics are necessary, but investigation of dynamic response needs the knowledge of mass distribution. For thermal analysis, three thermal parameters must be used: specific heat, coefficients of thermal expansion, and thermal conductivity data (the last two are defined as 3×3 matrices).

More data for structural analysis are elastic and strength characteristics of the lamina (laminas in case of hybrids, sandwich cores, or different thicknesses), number and angles of the laminas with respect to the laminate reference system, constraints, boundary conditions, loads, and so on, depending on the typology of the analysis (thermal, dynamic, etc.).

At the end, required results are obtained as requested by the designers (such as stresses, displacements, deformations, eigenvalues, eigenvectors, and so on. In the case of static structural analysis, for instance, stresses and deformations are described for all the nodes.

In this field, design criteria play an important role to determine the margin of safety in calculations. Many criteria have been proposed:

- Maximum stress
- Maximum deformation
- Tsai–Hill
- Tsai–Wu
- Tsai–Hahn
- Hoffmann and so on, including many other criteria that emphasize the role of interlaminar stresses in failure

It is recommended to use different criteria for the same structure, because the results can be sensibly different. The usual procedure is to accept the most conservative criterion.

The codes run several times with better refined mesh, following certain criteria. The procedure will continue until the advantages with respect to the previous configuration are less than 2% to 3%. Even a simple description, similar to the given one, is enough to give an idea of the complexity of the process. More details of the finite element method for static analysis are provided in the literature [9]. Finite element modelling of braided structures and composites has been discussed in detail in chapter 3 and 5 of this book.

Usually, local designs oriented to peculiar parts are necessary. Particular attention must be given to the junctions among different substructures. Junctions can be performed by welding or bonding or by the use of mechanical fasteners (bolts, rivets, screws, etc.).

After that, a configuration is stated, in the sense that requirements and specifications are established. At this point, it is necessary to organize schedule and resources to be used for the production of the first prototype (components and integrated structure). So, the testing phase begins for subcomponents and the assembled structure to validate the calculations. At the beginning, of course, nondestructive testing (NDT) is performed to check the quality of the piece. NDT is performed on the surface (visual inspections, penetrating fluids, and magnetic particles) and for the detection of internal defects (x-rays, ultrasonic inspections, and thermography).

After that, it could be useful or necessary to destroy some pieces to perform destructive inspections to analyze impact, aging, fatigue, thermal stability, and so on. Moreover, during its working life an aeronautical part is periodically checked, as described in the maintenance manual.

The fatigue tests play a special role; it is not only the single parts that have to be calculated and tested by accelerated procedures (50 and 10 Hz for isotropic and orthotropic structures) but also the full body (e.g., wings or fuselages) has to be tested following specialized standards to demonstrate its life.

Also, the entire structure of an aircraft must be subjected to a very complex fatigue test. The body is suspended on a complex structure to simulate the absence of ground reactions, in order to simulate the flight condition. Dummy structures, such as the engines, are mounted on the body. Fatigue loads are applied and distributed along the full body by the use of hydraulic actuators, every typology with its specific law and frequency. The whole testing system must be controlled by a computer (Figure 8.21). The testing time and costs are usually very high.

International standards (ASTM, MIL, EASA, etc.) have to be followed at every step of the design. So, base materials, tests on materials and subcomponents, structural functionality tests, and flight tests must be checked by the authority during the whole prototype development process, thanks to meetings called audits.

Authority personnel has to follow the design from the beginning: computer codes, base materials, testing procedures and test numbers, prototype assembly, products qualification tests, and so on. Further, each step is controlled.

Usually after several design reviews, the production of commercial pieces can start meeting the necessary standards of quality (Figure 8.22). In the case of spacecraft, the pieces successfully subjected to tests cannot be employed for practical use; only new, untested pieces can be used in launchers and satellites. It is easy to understand why costs and prices are so high in the aerospace field.

Last but not least, it is necessary to discuss about the disposal techniques. This phase opens a totally new scenario; in fact, to degrade glass fibers and carbon fibers processes at very high temperatures (even 3000°K), producing toxic gasses

FIGURE 8.21   Boeing 787 body, ready for the fatigue test.

System development life cycle and review evolution

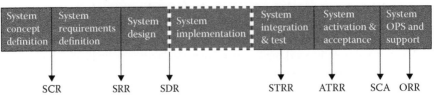

SCR = System concept review
SRR = System requirements review
SDR = System design review
STRR = System test readiness review
ATRR = Acceptance test readiness review
SCA = System configuration audit
ORR = Operational readiness review

FIGURE 8.22 System development life cycle and review evolution.

and eating up money because of the complexity of the processes, are necessary. So, in today's world the use of more environmental friendly materials such as green composites produced using natural plant fibres (such as hemp, flax, jute, and so on) and natural polymer matrices are being highly encouraged. Degradation of these composites is very cheap due to their natural properties. Such materials present low specific weight, good thermal and acoustic insulation characteristics, and excellent resistance to impact loads, better than glass fiber composite materials, but inferior to carbon fibers.

For space systems, disposal is necessary, especially for satellites that orbit around the Earth, to avoid the creation of space debris. So, satellites must have a sealed tank with the necessary fuel to perform deorbiting/parking maneuvers. These maneuvers are conceived as follows:

- Geostationary-orbit satellites have to free their spot within 25 years, going to a farther orbit called parking orbit and closing solar panels and antennas.
- Low-Earth-orbit satellites fall to Earth autonomously if they have a starting orbit lower than 500 km; otherwise, a sealed tank is needed.
- Medium-Earth-orbit satellites can go to a parking orbit, but it will cost a lot of money, or can be destroyed in the atmosphere by performing a deorbiting maneuver.

8.6 COMMERCIAL APPLICATIONS

The most interesting commercial parts are the following described parts for turbojets and fuselages:

- FJ44-4A turbofan engines, produced by Williams International and now flying on the Cessna Citation CJ4 business aircraft, are one of the first applications of braided fan cases on a commercial engine, thanks to the previously mentioned collaboration between NASA and A&P Technology [10,11].

- GEnx turbofan engines (Figure 8.23), produced by GE aviation, feature in the Boeing 747-8 (the new advanced version of the famous Jumbo Jet, extended and in the phase of redesign) and the Boeing 787. These engines are the first turbojets to have both fan case and fan blades in composite materials. In particular, the fan case is a braided structure, resulting from more than 20 years of research and development with NASA's partnership [12].
- Honeywell stator vanes made with A&P Technology's braided performs [5].
- New Airbus A350 fuselage frames [13,14], shown in Figure 8.24. Using highly automated circular machinery, EADS and SLG Kumpers had worked for 10 years to produce quasi-isotropic braided frames, sized to the aircraft fuselage diameter. The innovative technology allows the manufacture of a non-crimp final product, as waviness and wrinkles of the fibers have been reduced with the addition of unidirectional fabrics in 0° and 90° angles.

FIGURE 8.23   GE turbofan with braided fan case.

FIGURE 8.24   Manufacture of A350 fuselage frames.

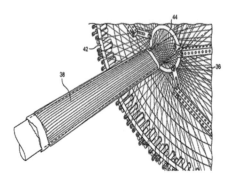

FIGURE 8.25   U.S. patent for braided helicopter blades construction.

8.7   RESEARCH STUDIES AND FUTURE TRENDS

As described earlier, the actual progress and stunning improvement in braiding was achieved only in the 1980s within the NASA ACT program. This program of partnership has been conducted by NASA along with some research institutions, such as Atlantic Research Corp.; some universities, such as Drexel University and University of Delaware; and the main U.S. aerospace companies McDonnel Douglas, Lockheed Martin, and Boeing. Since then, braided structures have already proved their efficiency in the aeronautical sector.

The future trends should be the improvement of textile machines to enhance the techniques of manufacturing of dry performs and their drapability to follow complex shapes. The use of braiding in compressed gas tanks, which would replace the filament winding technique by providing increased structural integrity and stability, is already under study and development.

Even non destructive inspections, such as ultrasonic controls and thermography, are under development to detect internal damage status. Moreover, high attention is being paid to the experimental determination of 3D elastic and mechanical characteristics, to be used also for the structural design software.

Regarding applications, some interesting ones could be considered in future in helicopter blades field (Figure 8.25) and for the construction of blades for high size wind turbines (especially offshore), considering high level of impact and fatigue loads, and also harsh environmental conditions.

ACKNOWLEDGMENTS

The author thanks his students Raffaele Gradini and Andrea Chiovini for their help in the bibliographic research and editing of this chapter.

REFERENCES

1. C. Scarponi, A.M. Perillo, L.Cutillo, "Advanced TTT composite materials for aeronautical purposes: CAI behavior," *Composites: Part B*, 2007, vol. 38, pp. (258–264).
2. L. Tong, A.P. Mouritz, M.K. Bannister, *3D Fibre Reinforced Polymer Composites*, Elsevier Science, Ltd., Oxford, United Kingdom, 2002, pp. (1–12, 22–32).

3. M. Brailey, M. Dingeldein, "Advancements in braided materials technology," A&P Technology, 2001, pp. (1–10).

4. D. Kelkar, J.D. Whitcomb, *Characterization and Structural Behavior of Braided Composites*, FAA report, Jan 2009, pp. (1–5, 1–6).

5. M. Dow, B. Dexter, *Development of Stitched, Braided and Woven Composite Structures in the ACT Program and at Langley Research Center*, NASA Langley Research Center, Hampton, Virginia, 1997, pp. (1–8, 12–31).

6. B. Griffiths, "Composite fan blade containment blade", Composite World, vol. 15, 2005, pp 1–4.

7. G. Roberts, M. Pereira, M. Braley, W. Arnold, J. Dorer, W. Watson, "Design and testing of braided composite fan case materials and components," NASA/TM, Oct 2009, pp. (1–11).

8. A&P technology, http://www.braider.com/Case-Studies/Bombardier-Wing-Flap.aspx, Access date: 6th May, 2015.

9. S. Mohajerjasbi, *Modeling and Analysis of 4-Step 3-D Cartesian Braided Composites*, 36th AIAA/ASME/ASCE/AHS/ASC New Orleans, 1995, pp. 8–16.

10. Textron Aviation, http://www.cj4.cessna.com/, Access date: 6th May, 2015.

11. Williams International, http://www.williams-int.com/, Access date: 6th May, 2015.

12. GE Aviation, http://www.geaviation.com/commercial/engines/genx/, Access date: 6th May, 2015.

13. G. Gardiner, "Airbus A350 Update: BRaF& FPP", Composite World, 2012, vol. 157, pp. 1–7.

14. A. Gessler, Textile technologies and preform manufacturing for advanced composites, SAMPE Swiss/technical conference of EAD IW, Jan 2001, Zürich, Switzerland.

9 Applications of Braided Structures in Transportation

Kadir Bilisik, Nesrin Sahbaz Karaduman,
and Nedim Erman Bilisik

CONTENTS

ABSTRACT

This chapter reviews the application of two-dimensional (2D) and three-dimensional (3D) braided structures in transportation industries, including their production methods and some properties. 2D and 3D braided preforms are classified based on various parameters depending on the yarn sets, yarn orientation and intertwining, micro-meso unit cells, and macro geometry. By using high strength and high modulus fibers, biaxial and triaxial 2D braided fabrics have been widely used as simple and complex shaped thin and thick (by over braiding) structural composite parts in transportation industries. However, delamination is always a basic failure mode in the over braided thick structure for transportation applications. For this reason, 3D braided fabrics are developed for transportation industry in which they have multiple layers and no delamination due to intertwine type out-of-plane interlacement. Recent multiaxis 3D braided fabrics have multiple layers and no delamination. The in-plane properties of multiaxis 3D braided fabrics may be enhanced due to the ±bias yarn layers. However, the multiaxis 3D braiding technique is at an early stage of development.

As per the discussions presented in this chapter, it can be concluded that the braided structures, both soft and rigid forms, are considered attractive materials for transportation industries due to their light weight and energy efficient potentials.

9.1   INTRODUCTION

Textile structural composites are widely used in various industrial sectors, such as civil and defense, as they possess some improved specific properties compared to basic materials such as metal and ceramics (Chou 1992; Dow and Dexter 1997; Kamiya et al. 2000; Ko and Chou 1989). Research conducted on textile structural composites has shown that they are delamination free and damage tolerant (Ko and Chou

1989). Two-dimensional (2D) biaxial, triaxial, and three-dimensional (3D) braided fabric structures are used as structural elements in various areas. For example, in transportation industries, they are used as plate, stiffened panels, beams, spars, and shell structures (Beyer et al. 2006; Donnet and Bansal 1990; Yamamoto and Hirokawa 1990). From a textile-processing viewpoint, 3D braiding is a preform technique used in the multidirectional near-net shape manufacturing of high damage tolerant structural composites (Ko 1987). 3D braiding is a highly automated and readily available manufacturing process. The fabrication of small-sectional 3D braided preforms is cost effective and not labor intensive (Dow and Dexter 1997). However, the fabrication of large sections of 3D braided preforms may not be feasible due to position displacement of the yarn carriers. A simple 3D braided preform consists of 2D biaxial fabrics and is stitched depending on stack sequence. Generally, 3D braided preforms are fabricated by traditional maypole braiding or innovative 4-step and 2-step braiding, or more recently by 3D rotary braiding and multistep braiding (Bilisik 2013; Kostar and Chou 1994a; Popper and McConnell 1987). Multistep braiding is a relatively new concept and with this technique, it is possible to make multidirectional 3D braided preforms by orienting the yarn in various directions in the preform (Bilisik 2013; Kostar and Chou 1994a). 3D braided preforms are classified based on various parameters. These parameters depend on the yarn type and formation, the number of yarn sets, yarn orientation and interlacements, micro-meso unit cells, and macro geometry. One of the general classification schemes was proposed by Ko and Chou (1989). Another classification scheme was proposed depending on micro-meso unit cells and macro geometry. In this scheme, 3D braided preform is divided into thin- and thick-walled tubes, which include a contoured shape and connectors, and special and mobile structures, which include structural holes and bifurcations (Lee 1990). Kamiya and Chou classified 3D braided structure based on manufacturing techniques as solid, 4-step, 2-step, and multistep (Kamiya et al. 2000). Bilisik (1991) proposed a more specific classification scheme of 3D braided preforms based on type of interlacement patterns, yarn orientation, and number of yarn sets. In the proposed classification scheme, as shown in Table 9.1, 3D braiding is divided into three categories as 3D braid, 3D axial braid, and multiaxis 3D braid that are non-interlaced inside but are interlaced on the outside of the preform surface. They are further subdivided based on reinforcement directions ranging from two to six with Cartesian or polar forms. This classification scheme may be useful for further studies on the development of multiaxis 3D braided fabric and 3D braiding techniques. This chapter reviews the application of 2D and 3D braided structures in transportation industries, including their production methods and some properties.

9.2 TYPES OF APPLICATIONS

9.2.1 Ropes

Braided ropes are made by 2D or 3D braiding processes where textile fibers are intertwined with each other. The properties of braided ropes generally depend on the type and linear density of the fibers and yarns, braid angle, directional fiber volume fraction, and pre- and postproduction heat and treatments. Simple braided ropes are produced by 2D maypole braiding. The bobbins are mounted on carriers that travel on

TABLE 9.1

The Classification of Three-Dimensional Braiding Based on Interlacement and Yarn Axis

Number of Yarn Sets	Three Dimensional Braiding					
	3D Braid		3D Axial Braid		Multiaxis 3D Braid	
	Cartesian	Polar	Cartesian	Polar	Cartesian	Polar

1 or 2	Square
	
	Rectangular
	• Through-the-thickness (Out-of-plane at an angle)
	• 1x1 pattern
	  2x1 pattern
	  3x1 pattern
	  4x1 pattern
		Tubular
		• Through-the-thickness (Out-of-plane at an angle)
		• 1x1 pattern
		  2x1 pattern
		  3x1 pattern
		  4x1 pattern
					Rectangular
					• Through-the-thickness (Out-of-plane at an angle)
						Tubular
						• Through-the-thickness (Out-of-plane at an angle)

3	Triaxial fabric • Braid yarn in surface (In-plane) Rectangular • Through-the-thickness (Out-of-plane at an angle) • 1x1 pattern, 2x1 pattern, 3x1 pattern, 4x1 pattern	Triaxial fabric • Braid yarn in surface (In-plane) Tubular • Through-the-thickness (Out-of-plane at an angle) • 1x1 pattern, 2x1 pattern, 3x1 pattern, 4x1 pattern	Rectangular • Through-the-thickness (Out-of-plane at an angle)	Tubular • Through-the-thickness (Out-of-plane at an angle)
4			Rectangular • Through-the-thickness (Out-of-plane at an angle)	Tubular • Through-the-thickness (Out-of-plane at an angle)
5 or 6			Rectangular • Through-the-thickness (Out-of-plane at an angle)	Tubular • Through-the-thickness (Out-of-plane at an angle)

Source: Bilisik, K., *Text. Res. J.*, 83, 1414–1436, 2013.

rotating horn gears, passing from one horn gear to the next. The horn gear is that part of the machine, which provides the rotating and intermittent motion based on braid patterns such as plain (1/1), twill (2/2), and diamond. Braided ropes are classified based on structural features as hollow braid, braid-on-braid rope, solid braid rope, parallel yarn rope, and kernmantle (McKenna et al. 2004). Some of them are explained in Sections 9.2.1.1 through 9.2.1.5.

Braided ropes are generally made from polyamide, polyester, polypropylene (PP), or high modulus polyethylene (HMPE), and para-aramid. Nylon is preferred for its strength and elastic stretch properties. However, nylon absorbs water and it gets weaker when it is wet. Polyester is as strong as nylon but stretches less under load and is not affected by water. It has somewhat better ultraviolet (UV) light resistance, and is more abrasion resistant. PP is preferred for its low cost and light weight (it floats on water) but it has limited resistance to UV light. It is also susceptible to friction and has poor heat resistance (http://en.wikipedia.org/wiki/Rope).

9.2.1.1 Hollow Braid Rope

Hollow braids are made in a maypole braiding machine, with half the carriers on the braider moving the strands clockwise and half moving them anticlockwise, while they also move in and out. The rope generally has eight or twelve strand braids, which are distinguished by having a hole down the center. A 96-mm diameter 12-strand single braid, which includes PP and polyester fibers, is shown in Figure 9.1a and b. Polyester offers good abrasion resistance while PP reduces weight and cost (McKenna et al. 2004).

9.2.1.2 Braid-on-Braid Rope

Braid-on-braid rope is made by traditional 2D maypole braiding consisting of a cover rope (sheath) over a braided core as seen in Figure 9.2a. The tension in the rope is shared by both the core and the cover. The core structure is produced with a long braid pitch. This provides high strength but relatively low extensibility compared to the cover. Jackets of low modulus fiber that do not support any load are often braided over a braided core of high modulus fiber to provide abrasion protection. Braid-on-braid rope is often selected because of their higher strength and lower degree of extensibility compared to hollow braid rope and parallel yarn rope (McKenna et al. 2004).

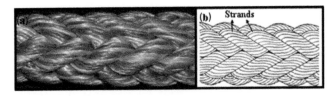

FIGURE 9.1 (a) 12-Single strand hollow braid, (b) Schematic view of hollow braid rope. (From McKenna et al., *Handbook of Fiber Rope Technology*, Woodhead Publishing and CRC Press, Cambridge, England, 2004.)

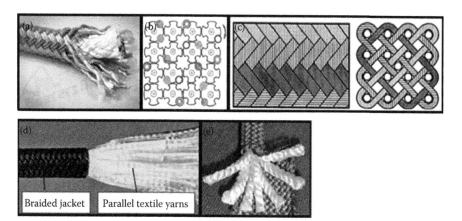

FIGURE 9.2 (a) Double braid rope, (b) Square braiding set up for solid braid rope, (c) Surface of solid braid rope and cross section of solid braid rope, (d) Parallel yarn braid rope (From McKenna et al., *Handbook of Fiber Rope Technology*, Woodhead Publishing and CRC Press, Cambridge, England, 2004.), (e) Kernmantle rope.

9.2.1.3 Solid Braid Rope

To make a solid braid rope, all the carriers on the braider move in the same direction. Each strand moves under another, moves to the side in the interior of the rope, comes back to the surface, and goes under again. All the strands on the surface appear aligned with the axis. Below the surface, they cross over to one side, always in the same direction. Figures 9.2b and c show schematically views of solid braid rope. A nylon core with polyester or multifilament PP can also be used to form solid braid rope (McKenna et al. 2004).

9.2.1.4 Parallel Yarn Rope

Parallel yarn ropes are low-twist ropes that have been developed to give high strength conversion efficiency of 80%–85% compared with 45%–60% for conventional braided ropes, and to minimize extensibility. Figure 9.2d shows actual views of parallel yarn braid rope. Fatigue life of this type of rope under cyclic tensile loading is excellent, exceeding conventional constructions and even wire rope, especially in a marine environment. On the other hand, parallel fiber ropes do not perform well in flexure load unless the radius is very large (McKenna et al. 2004).

9.2.1.5 Kernmantle Rope

Kernmantle rope consists of a core or "kern," which may be made of parallel multiple S and Z twisted yarns or multiple long pitch braids as shown in Figure 9.2e. This core is covered by a thin and twisted yarn braided jacket. Static kernmantle ropes have relatively low extensibility. They are used for rescue work and applications that do not require high energy absorption. In contrast, dynamic kernmantle is used for fall arrest, as would be experienced in mountaineering (McKenna et al. 2004).

9.2.2 CABLES

Cable consists of a number of ends that are braided to obtain an integrated structure. In mechanics, cables, as wire ropes, are used for lifting, hauling, and towing or conveying force through tension. In electrical engineering, cables are used to carry electric currents. An optical cable contains one or more optical fibers in a protective outer jacket that supports the fibers. Electric cables are mainly used in transportation, buildings, and industrial sites to transmit power. Thus, braided electrical cables have been extensively reported. An electrical cable has one or more conductors with their own insulations and individual coverings. They may be made more flexible by stranding the wires and braiding them together to produce larger wires that are more flexible than solid wires of similar size. Flexible cables used in transfer applications within cable carriers can be secured using cable ties. At high frequencies, current tends to run along the surface of the conductor. This is known as the skin effect. A cable radiates an electromagnetic field and picks up energy from any existing electromagnetic field around it. These effects are the result of an unwanted transmission of energy and may adversely affect nearby equipment or other parts of the same equipment. Therefore, there are special cable designs that minimize electromagnetic pick-up and transmission. One such cable design includes braiding of the cable strands together and shields them by a wire jacket. Figures 9.3a through d show some braided electrical cables (http://en.wikipedia.org/wiki/Cable).

In cable applications, metallic yarns are braided around the core material that carries the signal. In this way, the core material is shielded against EMI. Furthermore, the outer braided metallic structure could be jacketed to improve shielding effectiveness. Also, to make fire-resistant braided cable, the insulation is treated with fire retardant materials, or noncombustible minerals. Flat braids made of many copper wires are also sometimes used for flexible electrical connections between large components. The numerous smaller wires comprising the braid are much more resistant to break under repeated motion and vibration than a cable of larger wires. A common example of this is the braided cable connecting a car battery's negative terminal to the metal chassis. For high performance passenger and racing cars, Advanced fuel and ignition systems (AFIS) has introduced a comprehensive new range of spiral core ignition wire sets that have very low resistance, namely 50 ohms per foot, as seen in Figure 9.3d. Their low resistance means that more electrical current is transmitted from the distributor or coil to fire the spark plugs while still maintaining radio frequency interference and electromagnetic interference (RFI/EMI)

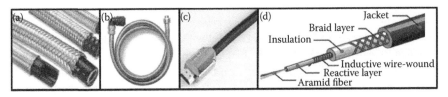

FIGURE 9.3 (a) Electrical wire protection cable (From www.diytrade.com, accessed on March 20, 2014.), (b) Braided flexible electrical conduit systems, (c) Metal HDMI cable (From http://china-hdmi-cable.en.made-inchina.com, accessed on March 25, 2014.), (d) AFIS 50 ohm wire. (From http://en.wikipedia.org/wiki/Cable, accessed on March 20, 2014.)

suppression. The electrical current is transmitted through para-aramide fiber of the cable and also its spiral core. At its center, the strong and heat-resistant synthetic core fibers are impregnated with carbon. The spiral core, which is wound around the fiber, prevents the magnetic field, the electrical noise, from being transmitted to other electrical components. In addition, the wire sets feature a superior jacket that promises the best combination of low and high temperature performance. The jacket is also resistant to dielectric failure, proving to be particularly effective if put in contact with under hood fluids—oil, brake or transmission fluids. Between the jacket and the spiral core, fiberglass braid is applied to improve terminal pull-off resistance (http://en.wikipedia.org/wiki/Cable).

9.2.3 Drive Shafts

A drive shaft must be capable of working continuously under conditions involving high speed and misalignment. In addition to transmitting torque and accommodating misalignments between the engine and transmission, the shaft running at high speed must weigh as little as possible to meet the natural frequency requirements. Composites offer significant advantages over metals for this type of application because of their lower weight, higher stiffness-to-weight ratio, higher fatigue strain capability and corrosion resistance (Faust et al. 1988). Drive shafts and flexible couplings are fabricated by using braiding technology and resin transfer molding (RTM). Thus, it results in a one-piece composite engine drive shaft that is a light, simple, more reliable and maintainable, and cost-effective system. A drive shaft has been fabricated by braiding on a 96-carrier braider using S-2 fiberglass yarn, as shown in Figure 9.4a. The braiding is done over a mandrel, which also acts as the tooling mandrel. A mold is used to cast the mandrel from a water-soluble casting compound. The mandrel is attached to a braiding take-up unit. The take-up unit is controlled by a microcomputer that is programmed to move the mandrel through the braider at the required speeds. A three layer braided drive shaft is achieved by over braiding. The braider's S-2 glass and carbon yarns are oriented between 30° and 60°. For effective design considerations, a small diameter shaft, and a large ratio of coupling's outside to inside diameter are desired for the couplings to operate at a fixed displacement

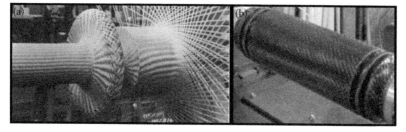

FIGURE 9.4 (a) 2D braided glass fiber preform for drive coupling shaft (From Ko, F.K., Braiding. In *Engineered Materials Handbook, Volume 1 Composites,* 519–528, Ohio, ASM International, 1987.), (b) 2D braided composite drive shaft. (From Faust et al., Development of an integral composite drive shaft and coupling, *National Technical Specialists Meeting on Advanced Rotorcraft Structures,* Williamsburg, Virginia, 1988.)

in the drive shaft. In addition, the natural frequency of the system is also one of the most important factors in the design of braided drive shafts. The material requirements for shaft coupling depend on thermo-mechanical loads and processing parameters. Figure 9.4b shows the 2D braided drive shaft composite structure. 3D finite element analysis (FEA) configured with the MSC/NASTRAN/PATRAN model has been used for the design of the integral shaft and coupling because of the orthotropic nature of the material (Faust et al. 1988).

9.2.4 STRUCTURAL COMPONENTS

In ground transportation applications, 2D and 3D braided structures should meet some general requirements such as low cost, manufacturability, good mechanical performance, no corrosion, reparability, and recyclability, as well as high damping, fuel economy, and low noise level. The energy absorption and structural integrity of braided structures highly prevent the components from delamination. Typical structural components in transportation engineering are knot elements for space frame-like structures, beams, shells, seats, and chassis. Because of the complex geometry and loading of these parts, cost-effective manufacturing techniques based on the 2D or 3D braiding processes are attractive. For instance, the use of braided composites in chassis, exhaust, and structural applications, as shown in Figures 9.5a through c, allows a significant reduction in component number and provides a substantial weight reduction compared with metals (Drechsler 1999). Furthermore, braided preforms and rigid composite connectors have been made by 2D or 3D circular braiding techniques enabling to fabricate the connector preform with multiple openings where connections are required, as shown in Figure 9.5d. This was achieved by directional intertwining where the particular section of the modular braiding carriers was rotated based on the structural opening part of the braided preform (Kurbak and Soydan 2010). 2D and 3D triaxial braids are more developed and widely applied than complex 3D braids. Coupled with the fully integrated nature and the unique

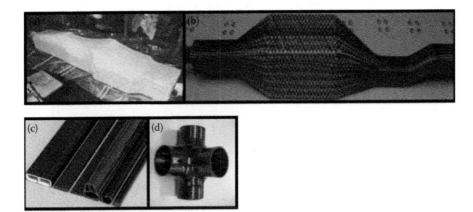

FIGURE 9.5 (a) Braided glass fiber complex geometry preform for car chassis, (b) Exhaust component for car, (c) Braided structural composites, (d) 3D braided composite connectors.

capability for near-net shape manufacturing, the current trend in braiding technology for structural applications includes the following: to expand to large-diameter braiding; to develop more sophisticated techniques for braiding over complex-shaped mandrels, multidirectional braiding with near-net shapes; and to extend the use of computer-aided design (CAD) and manufacturing.

9.2.5 Others

9.2.5.1 Battery Separators

Batteries have a separator to prevent short circuits that can occur between the anode and cathode. There are two types of battery: primary cells that cannot be recharged, and secondary cells that can be recharged (Bullock and Pierson 1992; Linden 1995). Electric lead–acid batteries used in motor cars consist of two plates, an anode and a cathode in dilute sulfuric acid. Polyester braided fabrics can be used for this type of application, which are stable under acidic conditions as shown in Figure 9.6a. The material must allow the flow of electrolyte but prevent actual particles from migrating. In addition, it must possess some degree of vibration resistance. The actual plates in car batteries are quite small, requiring pieces of braided fabric measuring about 15 cm by 15 cm but in forklift trucks or other electric vehicles they can be much larger. It was also reported that a full-scale battery box was braided and it was consolidated by RTM or vacuum assisted resin transfer molding techniques (Ko 2008).

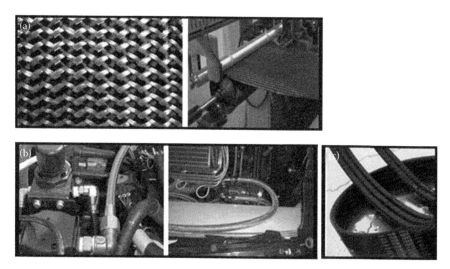

FIGURE 9.6 (a) Braided fabric for battery separator (From http://www.braider.com, accessed on September 20, 2014.), (b) Braided hoses for automobile engine connector (From http://www.wanderingtrail.com/Mods/JKmods/JKAiROCK/AiROCKTANKLINES.html, accessed on September 20, 2014.), (c) Braided hoses for automobile engine connector (From http://www.directindustry.com/prod/tempo-international/hydraulic-hoses-rubber-wire -braided-78801-743861.html, accessed on September 20, 2014.)

9.2.5.2 Hoses

In a study, braided hoses in vehicle applications demonstrated high performance under severe service conditions, and also they had low weight and required little space (Janssen 1996; Lambillotte 1989). Automotive hose products include fuel, oil, radiator heaters, hydraulic brakes, power steering, automatic transmission, and air conditioning pipes, as shown in Figure 9.6b (Mukhopadhyay and Partridge 1999). Generally, polyester and para-aramide fibers are used. Polyamide fiber is not commonly used in hoses because of its high extensibility but this specific property is useful in the expanding part of power steering hoses. It was reported that these items are very small but their reliability is extremely important (Scott 1971). Depending on the type of application, the ratio of the nominal working pressure to the actual bursting pressure of the particular hose is considered as a safety factor. The braiding angle is critical in controlling the behavior of the hose under pressure. If the helix angle chosen is approximately 54°, there will be a minimal change in the hose dimensions when pressure is applied. However, if the angle is lower than this, the hose will tend to expand and shorten under pressure. On the other hand, if the angle is greater than 54°, it will lengthen. For higher rated burst strength hoses, over braiding may subsequently be applied (Wootton 2001). Petrol pumps are usually fitted with special braided hoses, with inner tubes of highly oil-resistant rubber such as high nitrile polymers. The outer part of the hose must combine oil resistant polymers with good aging and ozone resistance, such as polychloroprene. The rubber is compounded to give good electrical conductivity and steel wires are incorporated, connected to the nozzle and the pump and through this to the earth. This prevents any buildup of static electricity, which could generate sparks with disastrous results (Wootton 2001).

9.2.5.3 Belts

Braided cords in rubber can be used in belts. 2D flat braided fabric structure can be made using high strength polyester fibers as shown in Figure 9.6c. For a long belt life, the braided structure must demonstrate high tensile strength, excellent flexural resistance, excellent shock resistance, and low extensibility. For instance, the toothed V-belt is shaped for maximum friction grip as well as high strength with compactness and rubber, usually chloroprene, is covered with a fabric/rubber jacket (Mukhopadhyay and Partridge 1999).

9.3 TYPES OF FIBERS

In braiding, generally high modulus and high strength fibers have been used due to high performance requirements. However, in recent years, cellulose-based natural fibers have been used in transportation industries, especially in the automobile sector due to their environmental friendly and recycling properties as well as their cost effectiveness. Both man-made and natural fibe based braided structures were utilized in ropes, cables, drive shafts, hoses, belts, and structural components

TABLE 9.2
Natural Fiber Properties

Natural Fibers	Density (g·cm⁻³)	Fiber Length (mm)	Fiber Diameter (µm)	Tensile Strength (MPa)	Tensile Modulus (GPa)	Failure Strain (%)	Moisture Regain 65% r.h. (%)	Applications
Cotton	1.55	20–64	11.5–17	300–700	6–10	6–8	8.5	Ropes Belts
Bamboo	1.50	2.7	10–40	575	27	—	—	—
Flax	1.4–1.5	27–36	17.8–21.6	500–900	50–70	1.3–3.3	12	Structural components
Hemp	1.4–1.5	8.3–14	17–23	310–750	30–60	2–4	12	Structural components
Jute	1.3–1.5	1.9–3.2	15.9–20.7	200–450	20–55	2–3	12	Rope Structural components
Sisal	1.3–1.5	1.8–3.1	18.3–23.7	80–840	9–22	2–14	11	—

Source: Cristaldi et al., Composites based on natural fibre fabric. In P.D. Dubrovski [ed.], *Woven Fabric Engineering*, 318–342, Sciyo Publishing, Croatia, 2010.

applications. Tables 9.2 and 9.3 show the specifications of some basic natural fibers and some high–low modulus/strength fibers, respectively. Seeds fiber (e.g., cotton) grows within a boll from developing seeds. Fruit fibers (e.g., coconut fiber) are extracted from the fruits of plant (www.fao.org/docrep/007/ad416e.htm). Cotton can absorb moisture up to 20% of its dry weight, and is also a good heat conductor. Bast fibers such as jute, flax, and hemp are found in the stems of the plants giving the plant its strength. Usually they run across the entire length of the stem and are therefore very long. Jute is a weak stem fiber due to its short fiber length. Flax is a strong fiber with a strength increase of 20% in wet conditions and it can absorb moisture up to 20% of its dry weight. Hemp fiber is strong and has the highest resistance to water among all natural fibers, but it should not be creased excessively to avoid breakage. Fibers extracted from the leaves are rough and sturdy and called leaf fibers, for example, sisal. When the high modulus and high strength fibers are considered, E-glass is produced on the largest scale followed by C-glass, D-glass, and S-2 glass fibers. Carbon fibers have a number of different varieties and their properties vary significantly depending on the conditions of manufacture. Most carbon fibers nowadays are made from acrylic fiber precursors. Para-aramide fiber has very high strength with temperature resistance, exhibiting 60% strength and modulus retention at 260°C. It does not melt but chars to a black color. Aramides are resistant to many solvents and have low water absorbency but they are sensitive to UV light and are not easily dyed. In addition, more specialist types of fibre have been produced including ceramic, boron, metallic, silicon carbide, and ultra HMPE fiber (Hearle 2001).

TABLE 9.3

Properties of Man-Made Fibers Used in Transportation

Type of Fiber	Density (g.cm^{-3})	Tensile Strength (MPa)	Tensile Modulus (GPa)	Failure Strain (%)	Melting Point (°C, d:degrade)	Moisture Regain 65% r.h. (%)	Resistance to Sunlight	Applications
Acrylic (High tenacity)	1.12–1.19	204–510	7.36	20–28	150[a]	1.3–2.5	Excellent	—
Nylon (High tenacity)	1.13	432–885	2.77	16–66	255	4–4.5	Poor	Ropes Belts Structural components
Polyester (High tenacity)	1.40	519–927	13.73	12–55	250	0.4–0.8	Good	Ropes Cables Belts Structural components Battery separator
Polyproplene (High tenacity)	0.91	318–675	7.06	20	165	0.1	Poor	Ropes Cables Hoses
Polyethylene (Ultra high modulus)	0.97	2590	117	10–45	115	None	—	Ropes Cables Hoses
Aramide	1.38–1.45	2900–3400	70–99	3.3–4	530[a]	6.5	—	Ropes Cables Belts Hoses Structural components

Material								Applications
Carbon	1.79–1.86	4000	230	1.8	3500^d	—	—	Drive shafts Structural components
Glass	2.50–2.70	2600	72	3	700	None	—	Cables Drive shafts Structural components Hoses
Polybenzimidazole	1.30–1.43	400	5.7	30	250^d	13	—	Structural components
Silicone Carbide	2.40–2.80	4000	420	0.6	1300	—	—	Structural and ablative component
Ceramic (Alumina)	3.60–3.90	1800	210	1.2	1760	—	—	Structural and ablative component
Steel	7.90	7600	150	4.8	1500	—	—	Hoses
Aluminum(6063)	2.70	100–245	68	14–27	660	—	—	—

Sources: McKenna et al., Handbook of Fiber Rope Technology, Woodhead Publishing and CRC Press, Cambridge, England, 2004; Hearle, J.W.S., High-Performance Fibers, Woodhead Publishing, Cambridge, England, 2001.

9.3.1 Polyamide Fiber

Polyamide fiber (nylon) molecules pack into a regular crystalline lattice. However, many techniques show that nylon fibers can be regarded as around 50% crystalline. The amorphous regions act like a rubber to give extensibility to nylon fibers, while the crystallites hold the structure together and limit the extensibility. In dry fibers at higher temperatures, there is freedom for tie-molecules to be extended, limited only by the connections to the crystallites, but, in dry fibers, hydrogen bonds between amide (–CO–NH–) groups cross-link the chains and stiffen the polymer. Increasing the degree of orientation increases the strength of nylon fibers. An advantage of nylon is its high elastic energy absorption, but its abrasion resistance is poor when wet (Hearle 2001; McKenna et al. 2004). Therefore, polyamide fiber is used for high tensile strength requiring areas such as various ropes in marine, industrial, climbing, and leisure activities; braided nets to make protective lines for petrol platform in the sea and marine functional applications.

9.3.2 Polyester Fiber

The most common type of polyester fiber is polyethylene-terephthalate, PET. In polyester fiber, the aromatic rings stiffen the structure in the amorphous regions. High-tenacity polyester yarns are melt-spun and drawn. Polyester fibers do not suffer from poor abrasion resistance. There are also other types of polyester fibers. Polytrimethylene-terephthalate and polybutylene-terephthalate contain three and four methylene (CH2) groups in the monomer respectively, instead of the two in PET. This gives greater flexibility to the chains and makes the fiber properties more like those of polyamide fibers (Hearle 2001; McKenna et al. 2004). Polyester fiber is also used in static climbing rope, mooring lines on small yachts, the mooring line rope for petrol platform as well as ropes for sailing and yachting, and marine functional applications.

9.3.3 Polyolefin Fibers

Polyolefins cover a range of fibers such as polyethylene (PE), PP, and their derivatives. PE has one of the simplest polymer chains, that is, (–CH2–)n whereas PP is made up of (–CH2–CH[CH3]–)n chains. The polyolefins are characterized by weak molecular interactions. Fibers are produced by melt-spinning and drawing. The resulting fibers are semicrystalline in structure. High modulus PE fibers are obtained by gel-spinning where the molecular chains are fully extended and oriented. They are commercially produced by Honeywell (Spectra™) and DSM (Dyneema™). PE is dissolved in a solvent at a fairly high concentration, extruded, and coagulated into a gel that can be super-drawn. This provides the required highly crystalline, highly oriented, and chain-extended structure. Interactions between the molecules are very weak and, in contrast to the liquid crystal fibers, its melting point is quite low (Hearle 2001; McKenna et al. 2004). Polyolefin fiber can be used in dynamic climbing ropes, ropes and netting for fishing industries as well as ropes for sailing and yachting.

9.3.4 ARAMIDE FIBERS

Aramide polymer, that is, polypara-phenylene-terephthalamide (PPTA) is basically a polyamide, but with –CO–NH– groups linked by benzene rings. The first fiber to be commercially produced was the para-aramide fiber, Kevlar, from DuPont. Twaron is another type of PPTA fiber. The PPTA polymer is expensive to make and to spin. It decomposes at around 450°C before it melts, and there are only few solvents that dissolve it. After polymerization, PPTA is dissolved in a concentrated sulfuric acid and then extruded through an air gap into a coagulating bath in a process known as dry-jet wet-spinning. PPTA molecules are relatively stiff and are strongly attracted to one another by the hydrogen bonding of –CO–NH– groups and phenylene (benzene ring) interactions. In solution, they form liquid crystal domains, which, after passing through the spinneret and being stretching in the air gap, become highly oriented. The resulting fiber has a fibrillar texture, but has the necessary highly crystalline, highly oriented, and chain-extended structure to give high modulus and tenacity. Additional processing under tension at elevated temperatures further straightens the chains and yields higher modulus forms of PPTA fibers. The molecules of PPTA are like flat ribbons. Along the chains, there are strong covalent bonds. Between molecules in the plane of the ribbons, there are moderately strong hydrogen bonding, but in the perpendicular direction, there are only weak van der Waals bonds, which are easily ruptured (Hearle 2001; McKenna et al. 2004). Technora, which is a fiber developed by Teijin differs in several ways from PPTA. Its copolymer forms an isotropic and not a liquid-crystal solution. The highly oriented and chain-extended structure results from the drawing and heat treatment after extrusion. An organic solvent is used and coagulation takes place in an aqueous bath. Among the high modulus fibers, Technora has a low creep, high melting point, and good abrasion and flexural resistance (Hearle 2001). Fully aromatic polyester molecules with some copolymerization can be melted and associated in liquid crystals as thermotropic liquid-crystal polymers (TLCP). Vectran is a polymer of this type that was produced by melt-spinning (Hearle 2001). Extensive research on rigid-rod polymers led to the development of poly(p- phenylene-2,6-benzobisoxazole) (PBO) fibers that are commercially produced by Toyobo as Zylon. Despite their high tensile strength, all the above high modulus-high tenacity fibers have weak transverse bonds, which cause them to have poor transverse properties, namely low shear modulus and strength, and a low compressive yield stress. As a result of this, M5 fiber was developed, which is known as poly (2,6-diimidazo[4,5-b 4′5′-e]pyridinylene-1,4-[2,5-dihydroxy]phenylene), or PIPD. The stronger intermolecular bonds in PIPD enhance the transverse properties (Hearle 2001).

Para-aramide fiber is used in clambing rope, the mooring line rope for petrol platform as well as ropes for marine functional applications. In addition, it is extensively used in composite forms in transportation as structural panels, car components, exhaust parts, joints, and connectors.

9.3.5 CARBON FIBERS

Carbon fibers are basically made of two forms: polyacrylonitrile (PAN) and pitch precursor. PAN-based fibers are made by a variety of methods such as wet-spinning or dry-spinning. Pitch-based fibers satisfy the needs of niche markets.

Vapor-grown fibers are entering commercial production, and carbon nanofibers seem to be promising for the future applications. Graphite fiber refers to a very specific structure, in which adjacent aromatic sheets overlap with one carbon atom at the center of each hexagon. This structure appears very rarely in carbon fibers even though they are conventionally called graphite fibers (Buckley and Edie 1993; Donnet and Bansal 1990). Carbon fiber is used in composite forms in transportation as drive shafts, structural panels, and various car components including tubes, joints, and connectors.

9.3.6 GLASS FIBERS

All glass fibers are derived from compositions containing silica. They exhibit useful bulk properties such as hardness, transparency, resistance to chemical attack, stability, and inertness, as well as desirable fiber properties such as strength, flexibility, and stiffness. Continuous glass fibers are made by a direct drawing process. Although still highly viscous, the resulting fibers are rapidly drawn to a fine diameter and solidified. Glass fibers fall into two categories, low-cost general-purpose fibers and premium special-purpose fibers. Over 90% of all glass fibers are general-purpose products namely E-glass. S-glass, D-glass, A-glass, ECR-glass, ultrapure silica fibers, hollow fibers, and trilobal fibers are examples of special-purpose glass fibers (Wallenberger et al. 2001). Glass fiber is used in composite forms in transportation industries as drive shafts, structural panels, and various components including tubes, joints, and connectors. In addition, it is used in electrical power cables and hoses.

9.3.7 CERAMIC FIBERS

Silicon carbide and alumina (aluminum oxide) are common ceramic fibers. Two standard commercial processes are utilized for producing ceramic fibers: chemical vapor deposition and spinning. In chemical vapor deposition, the ceramic is deposited onto a heated tungsten or carbon filament, through the introduction of a gas or gas mixture, containing silicon atoms. In the spinning technique, a precursor polymer is spun into filaments, whereupon the filaments are heated to convert them to ceramic fiber (Wallenberger 1999). Silicon carbide monofilaments can be produced by chemical vapor deposition onto a tungsten or carbon filament. Alumina fibers are produced by solution spinning and subsequent heat treatment. Due to their outstanding resistance to very high temperatures, ceramic fibers are primarily used in materials that require to withstand these temperatures. Braided fabrics made from ceramic fibers are used for the filtration of gases at high temperatures (Wallenberger 1999).

Ceramic fiber is used in composite forms in transportation as ablative structural parts, various components including rods, cones, tubes, and connectors.

9.4 TYPES OF BRAIDED STRUCTURES

9.4.1 2D BRAIDED FABRIC

9.4.1.1 Biaxial Braided Fabric

2D braided fabric is the most widely used material in industrial textiles, especially in the composite industry. It has one yarn set, namely braiders (oriented at an angle in $\pm\theta$ directions) which are intertwined with each other to form the braided fabric surface, as shown schematically in Figure 9.7a. Basically the fabric consists of diamond, regular, and hercules braid patterns, which are produced by traditional braid techniques (Brunnschweiler 1953). 2D braided fabrics can be layered according to the required thickness and consolidated to fabricate a rigid composite. However, it suffers from poor impact resistance because of crimp and low delamination strength due to the lack of binder fibers (Z fibers) in the thickness direction (Chou 1992). Although 2D layered stitched braided preform eliminates delamination weakness, it can reduce in-plane properties. Akiyama developed a braided fabric that consists of only braider yarns (±bias directions) as shown in Figure 9.7b (Akiyama et al. 1995). Head developed a tubular braided fabric that has two types of braider yarns (Head 2000). They are high modulus large tow size multifilaments and elastic small tow size multifilaments as shown in Figure 9.7c. Biaxial lace braided fabric (Branscomb et al. 2013) has two sets of yarns that intertwine with each other in the bias direction and the result is an open lace structure that provides minimum weight as seen in Figure 9.7d.

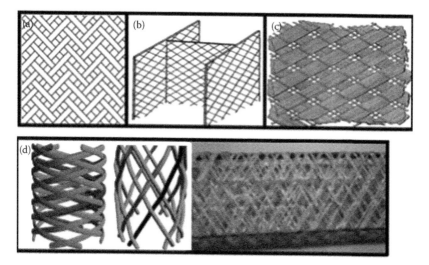

FIGURE 9.7   (a) 2D traditional biaxial braided fabric, (b) I-sectional braided fabric, (c) 2D non-interlace braided fabric, (d) Biaxial lace braid schematic and fabric. (From Bilisik, K., *Text. Res. J.*, 83, 1414–1436, 2013; Branscomb et al., *J. Eng. Fibers Fabr.*, 8, 11–24, 2013.)

9.4.1.2 Triaxial Braided Fabrics

Triaxial braided fabrics have basically three sets of yarns: +braid, -braid and warp (axial). Braided yarns intertwine with each other around the axial yarns at about 45° angle whereas axial yarns lie throughout the structure. As a result, the triaxial braided fabric is formed as shown in Figure 9.8a. This type of braided fabrics generally has a large open area between the axial yarns at the intertwining regions. Dense fabrics can also be produced with this technique. However, it may not be braided in a very dense structure compared to traditional biaxial braided fabrics. It was observed that the axial directional properties of triaxial braided fabric were superior to those of biaxial braided fabrics (Rogers and Crist 1997). Daimler Chrysler AG (now Daimler AG) developed an extra large 2D circular triaxial braiding machine with a robotic arm (Brandt et al. 2001), as shown in Figure 9.8b. Fiber Innovation Inc. also developed a large diameter circular 2D triaxial braiding machine as shown in Figure 9.8c. It should be noted that braided fabrics can be cut and stitched to make complex contoured shapes as done in the traditional garment industry (Fiber innovations Inc. 2002). Uozumi developed a 2D circular triaxial braiding process to combine braiding and curing to make on-line consolidated braided composites. The process is called the braiding-pultrusion technique (Uozumi et al. 2001). Uozumi also developed a technique to make various structural preforms for the aerospace industry (Hamada et al. 1995). After the 2D triaxial circular braided fabric is manufactured, it is deformed to the required cross section and stitched to provide interlaminar strength, as shown in Figure 9.8d.

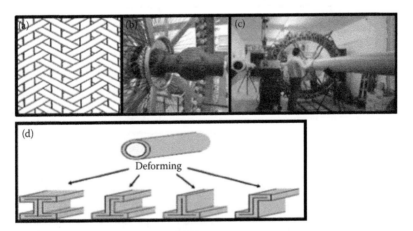

FIGURE 9.8 (a) 2D triaxial braided fabric, (b) 2D triaxial braiding machine by Daimler AG, (c) 2D triaxial braiding machine by Fiber Innovation, (d) Schematical view of 2D triaxial braided tube before and after deformation making sectional structural shape. (From Bilisik, K., *Text. Res. J.*, 83, 1414–1436, 2013.)

9.4.2 3D BRAIDED FABRIC

9.4.2.1 3D Fully Braided Fabric

There is one set of longitudinal yarns arranged in column and row directions in the cross section. All these yarns are intertwined with each other by at least four distinct motions in each machine cycle. The braider carriers move simultaneously in predetermined paths relative to each other within the matrix to intertwine the braiding yarns in order to form the braided preform. Florentine developed a 3D braided preform as well as its production method (Bilisik and Sahbaz 2012; Florentine 1982). The preform has layers and the yarns are intertwined with each other according to a predetermined path. In this way, the yarn passes the thickness (through-the-thickness) of the fabric and is biased such that the width of the fabric is at an angle between 10° and 70°, as shown in Figure 9.9a through d. Brown developed a 3D circular braided fabric (Brown 1988). The fabric has one set of yarns. The process has concentric rings connected to a common axis. Braid carriers are circumferentially mounted to the inside diameter of the ring. The ring is arranged side by side according to preform thickness. The rings rotate according to a predetermined path at only one braid carrier distance. Then, the braid carriers are shifted in the axial direction as shown in Figure 9.10a through c. Tsuzuki designed a 3D braider consisting of star-shaped rotors arranged in a matrix of multiple rows and columns (Tsuzuki et al. 1991). Four yarn carriers can surround a rotor and move in four diagonal directions that are determined by the rotation of the rotors, as shown in Figure 9.11a through c.

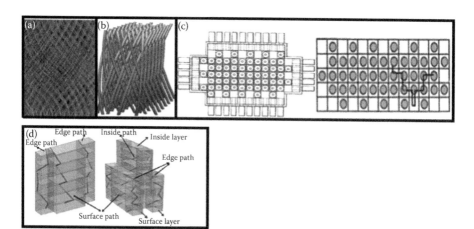

FIGURE 9.9 (a) 3D braided representative by 4-step method, (b) Unit cell of braided preform, (c) Schematical views of 3D braiding machine, yarn carrier path, (d) Braider yarn path on the edge and inside of the 3D representative braided preform with 4 layers and 6 layers, respectively. (From Bilisik, K. and N. Sahbaz, *Text. Res. J.*, 82, 220–241, 2012.)

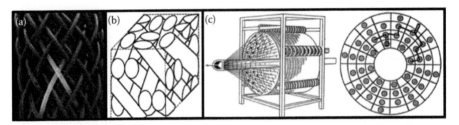

FIGURE 9.10 (a) 3D circular representative braided preform by 4-step method, (b) Unit cell of braided preform, (c) Schematical views of 3D circular braiding machine and yarn carrier path, respectively. (From Bilisik, K. and N. Sahbaz, *Text. Res. J.*, 82, 220–241, 2012.)

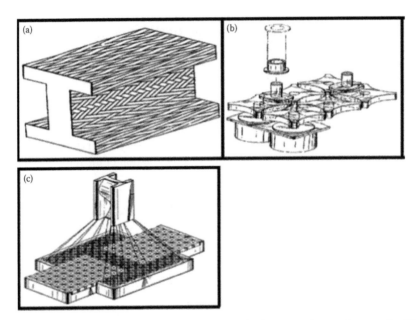

FIGURE 9.11 (a) Schematical views of 3D braided I-beam preform and rotary braiding machine, (b) Yarn carrier actuation unit, (c) Machine bed.

9.4.2.2 3D Axial Braided Fabric

A 3D circular axial braided structure can be formed by the maypole technique, which consists of two sets of yarns, warp (axial), and braid. The braider yarns are intertwined with the fixed axial yarns moving backwards and forwards radially around circumferential paths. Uozumi developed a 3D circular braided fabric that has ±bias (braider) and warp (axial) yarns (Uozumi 1995). Thick and various sectional fabrics, especially structural joint, end-fitting, and flange tube, were made by over braiding as shown in Figure 9.12a and b (Uozumi et al. 2001). Brookstein developed a tubular fabric that consists of braiders (±bias yarns) and warp (axial) yarns. Braiders intertwine around each axial yarn so that they lock each individual axial yarn in its place. This intertwining forms a helix structure, as shown in Figure 9.12c. The braiders are actuated

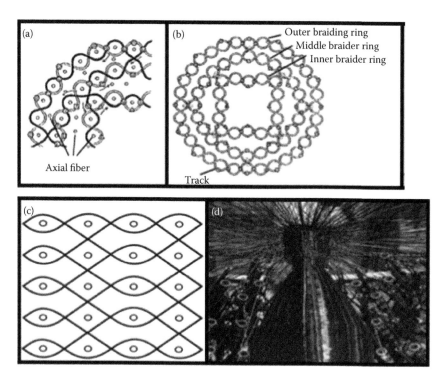

FIGURE 9.12 (a) Sectional schematic views of machine bed in maypole method, (b) Complete machine bed to show predetermined yarn path, (c) Unit cell of the 3D braided preform, (d) The braiding process in maypole method. (From Bilisik, K., *Text. Res. J.*, 83, 1414–1436, 2013.)

by a horn-gear mechanism to move in a predetermined path around the axial yarn, as shown in Figure 9.12d (Brookstein et el. 1994). To make a 3D braided flat preform in a 1 × 1 braid pattern, the braider carrier and axial yarns are arranged in a matrix of rows and columns. The first step is the sequential and reversal movement of the braider carriers in the column direction. The second step is the sequential and reversal movement of the braider carriers that is placed on the rapier in the row direction. The third step is again sequential and reversal movement of the braider carriers in the column direction. The fourth step is again sequential and the reversal movement of the braider carriers placed on the rapier in the row direction. These steps are repeated depending on preform length requirements. The representative 3D axial braided preform and unit cell are shown in Figure 9.13a and b (Bilisik 2011). In 3D flat axial braided prefom from the 2-step braiding process, axial yarns are arranged in a matrix array based on the sectional geometry of the braided structure. The braider yarns move along alternating diagonals of the axial array and interlock the axial yarns and hold them in the desired shape. The arrangement of yarns provides directional reinforcement and structural shape with a relatively small number of braider yarns. It was also shown that a variety of braided preforms including T-shape, H-shape, double T-shape (TT), and braided bifurcation preforms can be fabricated for using in the transportation applications. Mc Connell and Popper developed a 3D axial braided fabric in which the preform has layered axial yarns. They are arranged according to a cross-sectional shape, and braided yarns pass through the opening of axial layers to the row and column directions of the arrangement, as shown in Figure 9.13c and d (McConnell and Popper 1988; Popper and McConnell 1987). Schneider and Schneider developed a method and machine to make a 3D braided fabric, which has multiple axial yarn networks and braider yarns (Langer et al. 2000). The method is called 3D rotary braiding, which is similar to Tsuzuki's rotor braiding. The machine consists of horn gears that have a flat row–column array and each horn gear is actuated by an individual servo control motor and equipped with a clutch–brake mechanism to control the step or rotation of each single horn gear, axial yarn guide, and braider carrier. It is capable of making various sectional braided preforms with a CAD tool, as shown in Figure 9.14a through c.

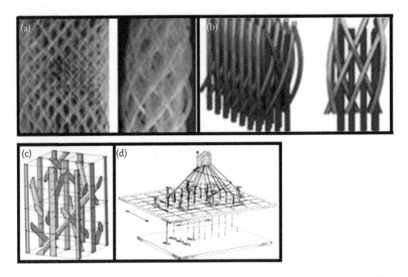

FIGURE 9.13 (a) 3D representative axial braided preform, (b) Unit cell, (c) 3D axial braided preform unit cell by 2-step braiding, (d) Schematic view of 3D axial braiding. (From Bilisik, K., *Text. Res. J.*, 81, 2095–2116, 2011.)

FIGURE 9.14 (a) Predetermined yarn path in 3D axial braided fabric by rotary braiding method, (b) 3D flat axial braiding machine, (c) 3D circular axial braiding machine. (From Bilisik, K., *Text. Res. J.*, 83, 1414–1436, 2013.)

9.4.2.3 Multiaxis 3D Braided Fabric

A multiaxial 3D braided structure produced by the 6-step method has ±braider yarns, warp (axial), filling, and Z yarns. The braider yarns are intertwined with the orthogonal yarn sets to form the multiaxis 3D braided preforms. The properties of the multiaxial 3D braided structure in the transverse direction are superior and the directional Poisson's ratios of the structure become identical. In this process, there are six distinct steps in each cycle. In steps 1 and 2, ±braider yarns are intertwined around the axial yarns as in the 4-step method. In step 3, filling yarns are inserted in the transverse direction. In steps 4 and 5, the ±braider yarns are intertwined around the axial yarns as in the 4-step method, and in step 6, Z yarns are inserted in the thickness directions (Chen and El-Shiekh 1994). Another multiaxial 3D braided structure produced by the 6-step method has ±bias yarns placed in the in-plane direction of the structure, and warp (axial), radial (Z yarns), and ±braider yarns placed in the out-of-plane direction of the structure (Bilisik 2013). The braider yarns are intertwined with the axial yarns whereas ±bias yarns are oriented at the surface of the structure and locked by the radial yarns to the other yarn sets. Figure 9.15a through d show the multiaxial cylindrical and conical para-aramide 3D braided structure produced by a 6-step method. Table 9.4 presents the specifications of multiaxial 3D braided Kevlar® preforms. The properties of the multiaxial 3D braided structure in the transverse direction could be enhanced and the nonuniformity in the directional Poisson's ratios of the structure could be decreased. In this process, there are six distinct steps in each cycle. In steps 1 and 2, ±braider yarns are intertwined around the axial yarns as in the 4-step method. In step 3, ±bias yarns are laid down on the surface of the structure. In step 4, the radial yarns move in the thickness direction of the structure and lock the ±bias yarns to the ±braider and axial yarns. In steps 5 and 6, the ±braider yarns are intertwined around the axial yarns as in the 4-step method (Bilisik 1998). Kostar and Chou developed a multistep braiding that is based on a computer algorithm (Kostar and Chou 1994b). In this way, the yarns make a large interlacement angle at the thickness of the fabric, which results in a large-sized unit cell. Also, the yarns may change more positions in the unit cells compared to the other 3D braiding processes (Kostar and Chou 1994a).

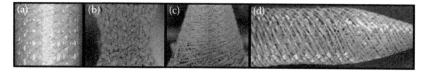

FIGURE 9.15 (a) Cylindrical multiaxis 3D braided para-aramide Kevlar® preform, (b) Tightly braided neck part of the conical Kevlar® preform, (c) Conical part of the Kevlar® preform, (d) Cylinder-conical Kevlar® preform. (From Bilisik, K., *Text. Res. J.*, 83, 1414–1436, 2013.)

TABLE 9.4

Multiaxis Three-Dimensional Braided Para-Aramide Preform by 6-Step Method

Fiber	Kevlar® 29(K29), Kevlar® 129(K129)
Axial yarn	1100 dtex (3 ply), K29
±Bias yarn	1100 dtex (4 ply), K29
Radial yarn	1100 dtex (1 ply), K129
±Braider yarn	1100 dtex (4 ply), K29
Structure	**Multiaxis 6-step 3D Braided Preform**
Axial yarn	2 (circumferential layers × 18 radial rows)
+Bias yarn	1 layer × 18 radial rows
−Bias yarn	1 layer × 18 radial rows
Radial	18 ends (one radial for every axial row)
Cross section	Cylinder
Dimensions	100 (outside diameter) × 5 (wall thickness) × 250 (length) mm
Preform tightness	Very high
Fiber	**Kevlar® 49**
Axial yarn	3400 dtex (3 ply)
±Bias yarn	3400 dtex (2 ply)
Radial yarn	3400 dtex (3 ply)
±Braider yarn	3400 dtex (2 ply)
Structure	**Multiaxis 8-step 3D Braided Preform**
Axial yarn	2 (circumferential layers × 18 radial rows)
+Bias yarn	1 layer × 18 radial rows
−Bias yarn	1 layer × 18 radial rows
Radial	18 ends (one radial for every axial row)
Cross section	Conical
Dimensions	140 (large diameter) × 55 (small diameter) × 5 (wall thickness) × 210 (length) mm
Preform tightness	Medium
Fiber	**Kevlar® 49**
Axial yarn	3400 dtex (3 ply)
±Bias yarn	3400 dtex (2 ply)
Radial yarn	3400 dtex (3 ply)
±Braider yarn	3400 dtex (2 ply)
Structure	**Multiaxis 6-step 3D Braided Preform**
Axial yarn	2 (circumferential layers × 18 radial rows)
+Bias yarn	1 layer × 18 radial rows
−Bias yarn	1 layer × 18 radial rows
Radial	18 ends (one radial for every axial row)
Cross section	Cylinder-conical
Dimensions	120 (large diameter) × 20 (small diameter) × 5 (wall thickness) × 440 (length) mm
Preform tightness	Low

Source: Bilisik, K., *Text. Res. J.*, 83, 1414–1436, 2013.

9.4.2.4 Comparison of Braided Fabrics and Methods

Chou and Ko explained that in 2D and 3D braided fabrics and composites, the design of the composite structural component depends mainly on loading conditions in end-uses (Ko and Chou 1989). In textile structural composite materials, the basic parameters are the properties of yarns which can be continuous monofilament or multifilament, matrix properties, total and directional volume fraction, preform architecture, yarn orientation in the architecture, and preform shape. These parameters together with end-use requirements determine preform manufacturing techniques. This requires sophisticated calculation techniques integrated with CAD/computer-aided manufacturing (CAM)–controlled preform and a composite manufacturing machine. Many calculation techniques have been developed with the aid of computer-supported numerical methods to predict stiffness and strength properties and to understand the complex failure mechanism of the textile structural composites (Chou 1992; Kamiya et al. 2000; Ko and Chou 1989). It was reported that 2D biaxial and triaxial braided fabrics have good balance in off-axis directions. They also have good drape to make complex parts. 2D braiding is an automated preform manufcturing process. However, 2D biaxial and triaxial braided preforms have limited thickness and sizes, and low out-of-plane properties. In 3D braiding, 3D braided preform has high out-of-plane properties and shows high drape to make complex parts during molding. It is also possible to fabricate near-net shape preforms. The process is a semiautomated one. However, it is slow and has size limitations. In addition, it is difficult to reproduce consistent preforms (Chou 1992; Kamiya et al. 2000; Ko and Chou 1989). Kamiya, Popper, and Chou compared 3D braided fabrics and methods based on the bias yarn placement, uniformity, the number of layers, and through-the-thickness reinforcements (Kamiya et al. 2000). It was concluded that 3D braided fabrics and methods are readily available. A more general comparison is carried out and presented in Table 9.5. As seen in the table, the 3D braided fabric parameters are yarn sets, intertwining method, yarn directions, preform shape, the number of layers, and fiber volume fraction. The 3D braiding process parameters are bed arrangement based on the predetermined yarn path, manufacturing type such as continuous or part, braider carrier type, and yarn volume in the carrier, packing, and the development stage. It can be seen that biaxial and triaxial braided fabrics and 3D braided fabrics are well developed and are commercially available. However, multiaxis 3D braided fabrics with additional yarn sets are still in the early stages of development (Bilisik 2013).

As seen in the discussion on 2D and 3D braided fabric structures, various thin and thick braided fabrics were developed based on end-user requirements. Table 9.6 outlines the braided fabric structures depending on their properties and possible specific applications in transport industries.

TABLE 9.5

Comparison of Three-Dimensional Braided Fabrics and Methods

Fabric	Yarn Sets	Method	Yarn Directions	Fabric Shape	Fiber Volume Fraction	Development Stage
Florentine (Florentine 1982)	One	4-step	±Bias (Out-of-plane at an angle)	Flat or complex shape (Multilayer)	Low or Medium	Commercial stage
Brown (Brown 1988)	One	4-step	±Bias (Out-of-plane at an angle)	Circular or complex shape (Multilayer)	Low or Medium	Commercial stage
McConnell and Popper (McConnell and Popper 1988)	Two	2-step	±Bias/Axial (Out-of-plane at an angle)	Flat or Complex shape (Multilayer)	Medium or High	Commercial stage
Brookstein (Brookstein et al. 1994)	Two	Maypole	±Bias/Axial (Out-of-plane at an angle)	Circular or Complex shape (Multilayer)	Low or Medium	Commercial stage
Tsuzuki (Tsuzuki et al. 1991)	One	Rotary	±Bias (Out-of-plane at an angle)	Flat or Complex shape (Multilayer)	Medium or High	Commercial stage
M. Schneider and H. Schneider (Langer et al. 2000)	Two	Rotary	±Bias/Axial (Out-of-plane at an angle)	Flat or Complex shape (Multilayer)	Medium or High	Commercial stage
Chen and El-Shiekh (Chen and El-Shiekh 1994)	Four	6-step	±Bias/Axial/ Filling/Z-yarn (Out-of-plane at an angle and orthogonal)	Flat or Complex shape (Multilayer)	Medium or High	Early prototype stage
Bilisik (Bilisik 2013; Bilisik 1998)	Four	6-step	±Bias (In-plane) ±Bias/Axial/Radial (Out-of-plane at an angle)	Circular or Complex shape (Multilayer)	Medium or High	Early prototype stage
Kostar and Chou (Kostar and Chou 1994b)	One	Multistep	±Bias (Out-of-plane at angle)	Flat or Complex shape (Multilayer)	Medium or High	Commercial stage

Source: Bilisik, K., *Text. Res. J.*, 83, 1414–1436, 2013.

TABLE 9.6
Comparison of Braided Fabric Structures

Braided Fabric Structures	Properties	Applications in Transportation	Specific Section
2D Braided Fabric			
Biaxial Braided Fabric	Enhanced off-axis strength and stiffness, good in-plane shear strength and modulus, better under uniaxial and biaxial loading, good torsion, good internal pressure, fair bending strength, good drapeability and shapeability	Ropes, cables, hoses, belts	Decorative ropes; marines, climbing ropes, mooring line ropes, connector hoses in automobile
Triaxial Braided Fabric	Enhanced axial and off-axis strength and stiffness, fair drapeability and shapeability, restricted jamming due to axial yarn and good dimensional stability; adding holes and joint part or connection during fabrication	Structural components	Interior component of train and automobile, side panel, body structures, roof panel
3D Braided Fabric (Through-the-Thickness)	Enhanced in-plane and out-of-plane strength and stiffness, good in-plane and out-of-plane shear strength and modulus, good torsion, good internal pressure, good fracture toughness, no delamination between layers, integrated layered structures, good drapeability and shapeability, adding holes and joint part or connection during fabrication	Structural composite components	Automotive pillar; rods, beams; racing car body, structural connector; foam filled parts, drive shafts, engine valve
3D Braided Fabric by Over Braiding	Enhanced in-plane and off-axis strength and stiffness, poor out-of-plane strength and stiffness, limited fracture toughness, delamination between layer; suitable for complex part geometries	Structural composite components	Interior component of train, truck, and automobile, side panel, body structures, structural engine components
3D Axial Braided Fabric (Through-the-Thickness)	Enhanced in-plane and out-of-plane strength and stiffness, good in-plane and out-of-plane axial tensile and shear strength and modulus, good torsion, good internal pressure, good fracture toughness, no delamination between layers, integrated layered structures, fair drapeability and shapeability, good bending and dimensional stability	Structural composite components	Automotive pillar; rods, beams; racing car body, structural connector; foam filled parts, drive shafts; engine valve, seats and chassis
3D Triaxial Braided Fabric by Over Braiding	Enhanced in-plane axial tensile and off-axis strength and stiffness, poor out-of-plane strength and stiffness, limited fracture toughness, delamination between layer, suitable for complex part geometries	Structural composite components	Interior component of train, truck and automobile, side panel, body structures, structural engine components, exhaust part, connector and joint parts
Multiaxis 3D Braided Fabric	Enhanced in-plane and out-of-plane strength and stiffness, good in-plane and out-of-plane multiaxial tensile and shear strength and modulus, good torsion, good internal pressure, good fracture toughness, no delamination between layers, integrated layered structures, fair drapeability and shapeability, good bending and dimensional stability	Structural composite components	Shell component for automotive, rods; beams, racing car body, structural connector, drive shafts, engine valve, seats and chassis, missile and rocket parts

9.5   PRODUCT DESIGNING AND EXAMPLES

9.5.1   ROPE

Primary requirements for a rope design are the high strength and stiffness. Therefore, rope breaking load must be greater than the maximum tension applied by a specified safety factor. Torque can lead to twisting of a rope against terminations, which could bring about damage and local failure. The predicted fatigue life before failure under imposed conditions must be within safety limits (McKenna et al. 2004). The breaking load and resistance to elongation (axial stiffness) of a rope clearly increase with increasing size of rope. However, there is a contraction in length during rope formation, because the components lie at an angle to the rope axis due to the braiding. Figure 9.16a through c show the schematic view of the rope structure design concept. The following relations can be used for a rope design (McKenna et al. 2004).

$$LD_r = N \times LD_y \times C \tag{9.1}$$

$$C = L_y / L_r \tag{9.2}$$

$$F_r = LD_r \times f \times CF \tag{9.3}$$

$$S_r = LD_r \times E_f \times CF_s \tag{9.4}$$

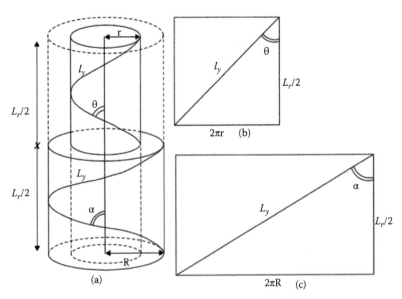

(a)

(b)

(c)

FIGURE 9.16   (a) Schematic views of braided rope structure, (b) Opening of the central part of the braided rope structure, (c) Opening of the outside part of the braided rope structure. (From McKenna et al., *Handbook of Fiber Rope Technology*, Woodhead Publishing and CRC Press, Cambridge, England, 2004.)

where, LD_r is the rope linear density; N is the number of yarns in rope; LD_y is the yarn linear density; and C is the contraction factor on rope structure; L_y is the average yarn length in the rope structure and L_r is the length of rope; F_r is the rope breaking load, N; f is the fiber tenacity, N/tex; and CF is the strength conversion efficiency ratio; S_r is the axial stiffness, N; E_f is the fiber modulus, N/tex; and CF_s is the stiffness conversion efficiency ratio.

The ultimate failure mode for a rope under load is creep rupture, which is called static fatigue. Cyclic loading results in other forms of damage and hysteresis heating. Abrasion damage that occurs during internal and external abrasion has been shown to be a serious problem in highly structured ropes. At low rope tensions, it is possible for some yarns to go into axial compression while the tension is taken by other yarns.

9.5.2 HOSE

Bursting pressure can be calculated for most braided hose types, based on the tensile strength and the method of reinforcement. Figure 9.17 shows the schematic view of the braided hose structure. Generally, the conversion efficiency (actual measured burst strength to the calculated values) works out at around 75%–80% for most hose constructions. As a general guide, hose working pressures are classified into three pressure ranges as low pressure below 2.0 MPa, medium pressure between 2.0 and 7.0 MPa, and high pressure over 7.0 MPa. Burst strength for braided hose can be represented as follows (Wootton 2001):

$$P = 0.2N_s R \sin\theta / DL \tag{9.5}$$

where, P is the bursting pressure, bar; θ is the braid angle, degree; N_s is the total number of yarns in the braided structure; R is the tensile strength of yarn, N; D is the average diameter of the braided structure, cm; L is the pitch length of the structure, cm.

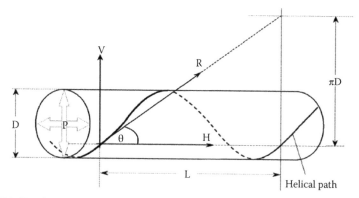

FIGURE 9.17   Schematic views of braided hoses structure. (From Wootton, D.B., *The Application of Textiles in Rubber*, 187–212, Rapra Technology, Exeter, United Kingdom, 2001.)

9.5.3 COMPOSITE COMPONENT DESIGN

Fiber orientation (θ) and volume fraction (V_f) are key engineering parameters for a braided textile composite from formability, permeability, and performance stand points. The manufacturing of composites often requires transformation of the fiber reinforcements into various structural shapes through net-shape fabrication or formed-shape processing. Although 3D textile preforming is more suitable for the creation of net structural shapes, 2D braided preforms are usually formed into shapes by molding, and if required, stitching. In addition, triaxial fabrics are more adaptable to 3D draw molding than biaxial fabrics. In quantifying the formability of fabrics, yarn slippage and low yarn jamming angles are required for fabric conformability. Accordingly, in fabric formability modeling, fiber volume fraction distribution, fiber orientation, and fiber interlacing intensity as well as the limit of geometric deformation must be considered. The fluid flow permeability of textiles is an indication of how easily and uniformly a matrix can be infiltrated into the fibrous assembly. The permeability of textile preforms is affected by the dynamic interaction of fiber architecture and fiber volume fraction. It was found that the introduction of through-the-thickness fibers significantly increases the permeability of the preforms, especially for preforms with high fiber volume fraction (Ko 2008).

The mechanical behavior of a composite depends on fiber orientation, fiber properties, fiber volume fraction, and matrix properties. The fiber volume fraction is related to the machine in terms of the number of yarns and the orientation of those yarns. The fiber geometry is also strongly related to the machine which determines the orientation of the fibers and the final shape. The shape is formed using a mandrel, and the fiber volume fraction can readily be determined from the orientation and amount of fibers used, as shown schematically in Figure 9.18a through c. The total material area of yarns in a given cross section of a composite preform can be determined as follows (Ko 2008):

$$A_m = A_y \times N_y / \cos \theta \tag{9.6}$$

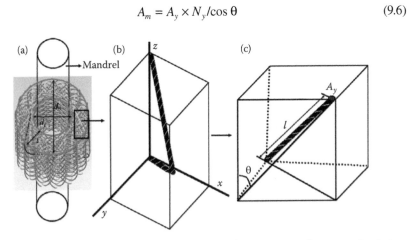

FIGURE 9.18   (a) Schematic views of 3D braided preform structure for composite design, (b) Unit cell of the 3D braided preform structure for composite design, (c) Defining of unit braided yarn in the unit cell. (From Ko, F.K., Braiding. In *Engineered Materials Handbook, Volume 1 Composites,* 519–528, Ohio, ASM International, 1987.)

where, A_m is the area of material in the cross section, mm²; A_y is the cross-sectional area of the yarn, mm²; N_y is the number of yarns on the machine (M × number of plies, where M is the number of carriers on machine); and θ is the orientation of the braider yarns with respect to the mandrel axis, degree. Thus once the composite dimensions are known, the fiber volume fraction can be expressed as follows:

$$V_f = A_m / A_c \qquad (9.7)$$

where V_f is the fiber volume fraction, %; A_c is the cross-sectional area of the composite, mm²; and A_m is the cross-sectional area of material in the composite, mm². If a composite of a given cross-sectional area and a particular yarn and fiber volume fraction are required, the fabric can be designed based on the number of plies and the orientation of the yarns. The analytic relation can be given as follows:

$$\cos \theta = MN_{ply} A_y / \left(V_f A_c \right) \qquad (9.8)$$

where, N_{ply} is the number of plies per bobbin. Thus, the design is determined for a certain numbers of plies. In summary, the braiding parameters for braided composites can be presented in the following equations as follows:

$$d_0 = MN_{ply} A_y / \left(\pi t V_f \cos \theta \right) + t \qquad (9.9)$$

$$d_i = MN_{ply} A_y / \left(\pi t V_f \cos \theta \right) - t \qquad (9.10)$$

where, d_0 is the outside braid diameter, mm; d_i is the inside braid diameter, mm; and t is the composite (fabric) thickness, mm. With this equation, the effect of braiding angle, fiber volume fraction, and the number of plies on the number of carriers required to produce a given braid diameter for a specific composite can be calculated. The take-up/rotation ratio (R) indicates the distance the mandrel traverses for one rotation of the carriers. Thus, for a given mandrel diameter (d) the relation between θ and R is given as follows:

$$d = R \tan \theta / \pi \qquad (9.11)$$

Thus, to maintain the proper fiber orientation (and thus the desired fiber volume fraction), the machine should be set for a take-up/rotation ratio of R. If the mandrel is of irregular shape, R can be monitored and modified accordingly along the length of the mandrel. Using these equations, one can easily determine the total number of yarns required to make a fabric with a given fiber volume fraction and cross-sectional area if the parameters like fiber density, yarn linear density, and yarn surface angle are known. The maximum volume fraction that is attainable with a given construction is dependent on the fiber architecture. The fabric geometry method (FGM) is a volume averaging homogenization method, which takes into consideration the volumetric angular distribution associated with fiber architecture via well-established coordinate transformation methodology. The input information generated from FGM is incorporated into the Algor (Algor Inc. 1997) FEA program, along with standard composite engineering practices, to optimize the laminate schedule and to minimize weight in the vehicle (Ko 2008). The pillars are the structural backbones of the top body shell of the electric vehicle. The requirements for high torsional and bending stiffness make triaxial braid a suitable candidate for the pillars. The required fiber orientation and fiber volume fraction to meet the design criteria as quantified in the laminate analysis are reduced to braiding machine control parameters such as braiding point location, braiding speed, and take-up speed for programming the braiding machine for the automated braiding process. For example, the pillar/seat-rail structure experiences a high seat belt load in a crash situation. To reduce the assembly time, a triaxial braided structure was designed and built by mapping the laminate properties to the braided composite properties generated by the FGM (Ko 2008). On the other hand, a feature of the braiding process, which makes it particularly interesting in composite applications, is the relative ease with which cutouts, pins, fasteners, and fittings can be incorporated into the work pieces. For example, if a transverse pin is positioned in a cylindrical mandrel the braiding yarns can accommodate the geometry of the pin automatically, and though they follow a locally distorted path there are no yarn ends in the vicinity of the pin and the full strength of the yarn assembly is maintained. If the pin is subsequently removed, a fully reinforced, integrally incorporated hole remains. The failure load in a pin-loaded hole is approximately 1.8 times greater for a braided hole than for a machined hole, and the tensile failure load of cylinders with braided holes is approximately 1.23 times that of cylinders with machined holes (Skelton 1989).

9.5.3.1 Pattern and Jamming in 3D Braided Fabric Design

In 3D braided fabric design for composite, the pattern and the jamming are the critical design parameters to be considered. For instance, the pattern, yarn interlacement frequencies in the 3D braided preform design, directly affects the unit cell dimensions in the fabric. In addition, it influences the braider yarn angle (θ) and predetermined yarn path, which relates to braider yarn length (l) in the preform structure. On the other hand, the jamming, deformation limits in the dry form of the braided structure, relates to the introduction of yarn sets in the braided preform and influences the shapeability and moldability of the braided preforms.

It was demonstrated that braid patterns influence the 3D braided and 3D axial braided unit cell structures produced by the 4-step method. The unit cell structure has a fine intertwine in the 1×1 pattern, whereas it has a coarse intertwine for other braid patterns such as 2×1, 3×1, or 4×1. This could affect the preform unit cell sizes. The braider angle (θ) of 3D braided preform slightly decreased when the braid pattern changed from 1×1 to 3×1 whereas the braider angle of 3D axial braiding slightly increased when the braid pattern changed from 1×1 to 3×1 (Bilisik 2011; Bilisik and Sahbaz 2012). Therefore, the pattern influences the composite volume fraction (V_f) and void content, braider yarn angle (θ), and yarn length (l) (Bilisik and Sahbaz 2012). Jamming conditions considerably influence 3D braided and 3D axial braided unit cell structures for all braid patterns. Minimum jamming decreases the width of the unit cell structures whereas maximum jamming increases their width. Width reduction of the unit cell structure in the 1×1 pattern was high compared to those of 2×1, 3×1, or 4×1 patterns. Also, minimum jamming increased the densities of the 3D braided and 3D axial braided unit cell structures whereas maximum jamming decreased their densities (Bilisik 2011; Bilisik and Sahbaz 2012).

9.6 COMMERCIAL APPLICATIONS

9.6.1 DEEP WATER MOORINGS

It has been reported that chain mooring systems lose effectiveness between 100 and 200 m of depths due to weight reduction. It was claimed that a composite system consisting of a double braid polyester line connected to relatively short chains, gives nearly equal performance at all depths and is superior to wire rope or chain at over 200 m of depths. In offshore platforms, steel wire cables are successfully used at up to 500 m of depths, but are unduly heavy for 1000 to 3000 m of depths. Fiber ropes offer an alternative, as seen in Figures 9.19a and b. The two principal design criteria, off set and peak load, are determined by the environmental forces that include oceanic surface waves as gravity pulling and capillary, mooring geometry, line lengths, and rope tensile properties. The fundamental rope properties used in deep water moorings are rope stiffness including post-installation and storm stiffness, axial compression fatigue, rope braid angle, bending rigidity, UV light and chemical resistance, and internal–external wear. Low-twist parallel strands and wire–rope constructions have been used in offshore moorings. Depending on wave heights, polyester and nylon ropes can be utilized at shallower depths. High modulus fiber ropes, which have lower weight, should be considered at greater depths, such as for the mooring of air defense platforms (http://taflab.berkeley.edu/; Mckenna et al. 2004).

FIGURE 9.19 Types of offshore structures based on operating water depth. (a) Some fiber rope mooring system (From http://taflab.berkeley.edu/ME168-FA13/ME168_Applications. htm, accessed on May 15, 2014.), (b) Fiber rope laying in deep sea water. (From http://www .offshoreenergytoday.com/emerson-to-upgrade-safety-and-automation-systems-on-visund-platform, accessed on May 6, 2014.)

9.6.2 Climbing Ropes

Early climbers used natural fiber ropes, and this continued through the advances in rock climbing. Nylon kernmantle ropes dominate the market for dynamic ropes, because their moderately high strength combined with high breaking extension gives high absorption of energy to arrest the fall of a climber. The comparatively low modulus prevents the applied forces from being too large, and ensures good elastic recovery, as shown in Figure 9.20a. A disadvantage of nylon is the loss of strength and work of rupture when wet, but the effect of water can be reduced by special coatings. Newly available polyester fiber that does not absorb moisture may prove useful for climbing ropes. High modulus grades of nylon 6.6 are used in fixed ropes used in caving, where high strength and low extensibility are required. Also, canyon rope uses a PP core (Meriam and Kraige 1978; Strorage and Ganter 2014).

9.6.3 Sailing and Yachting

The demands on ropes in this area are low stretch, high strength, and a long service life. Fiber ropes have always been used for sheets, because of continuous handling, but they are now increasingly being replaced by steel wire ropes for halyards. Surface frictional and wear characteristics are important for lines used on winches (capstans). Important features are good grip when held under tension with a minimum number of wraps and smooth sliding when tension is released. Vectran rope, which is described as the ultimate sheet/halyard for racing and large yachts and ideal for hydraulically tensioned halyards and for backstays, has 12 strands and is designed for exceptional abrasion resistance while maintaining flexibility and elongation that is equal to steel wire rope. PE rope coated with polyurethane exceeds the strength of other high-performance fiber ropes and it has a strength/weight ratio eight times that of steel wire rope as shown in Figure 9.20b (Belgrano and O'Connell 1992).

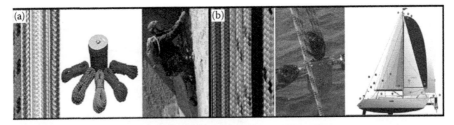

FIGURE 9.20 (a) Braided rope for climbing application (From http://www.pelicanrope. com /static_kermantle_roeps.html, http://sicksport.com/rock-climbing-climbing-rope-c-48_78 /climbing-rope-lm-dynamic-ropes-p-1097, http://www.ropeinc.com/climbing-safety-dynamic -ropes.html, accessed on October 1, 2014.), (b) Braided rope for sailing and yachting application (From http://nhmarine.com.au/rope/double-braid-yacht-rope-polyester.html, http:// www.nauticexpo.com/prod/baltic-ropes-magistr/yacht-dock-lines-polyester-core-polyester -sheath-33266-244907.html, accessed on October 1, 2014.).

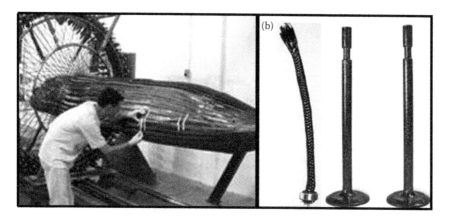

FIGURE 9.21 (a) Carbon fiber braided structure for marine application (From http://www
.compositesworld.com/articles/autoclave-quality-outside-the-autoclave, accessed on October
2, 2014.), (b) Carbon fiber based braided structure for engine valve application (From http://
www.compositesworld.com/articles/composite-engine-valves, accessed on October 2, 2014.).

9.6.4 Marine

Braided composites can be used in minesweepers, sonar domes, cargo ships, patrol
and pleasure boats. In addition, composites are being increasingly used for naviga-
tional aids such as buoys. Figure 9.21a shows the carbon fiber braided preform fabrication for the marine structural part. Recently, a new generation hovercraft has been
designed, which makes use of aramide braided composites in place of aluminum.
Its advantages include lighter weight, corrosion resistance, less noise in operation,
better shock absorbency, and higher abrasion resistance to rocks and sand surfaces.
In addition, all marine vessels use large amounts of braided material for vibration,
thermal, and noise insulation, especially in and around turbines and engine rooms
(Summerscales 1987).

9.6.5 Automative

2D and 3D braided preform and composites have been used in racing car bodies,
structural members such as beams that are made up of foam cores over braided with
a carbon preform structure, aprons and spoilers, and connecting rods, as shown in
Figure 9.21b (Popper 1990). Also, car noses, monocoques, and bumpers are made
from braided carbon structures. They reduce weight and improve the crash behavior
(Gries et al. 2008).

9.7 RESEARCH STUDIES AND FUTURE TRENDS

Biaxial and triaxial 2D braided fabrics have been widely used as simple- and
complex-shaped structural composite parts in various technical areas. In addition,
biaxial and triaxial braiding methods and techniques are well developed. However,
2D braided fabrics have size and thickness limitations. 3D braided fabrics have

multiple layers and show no delamination due to intertwine type out-of-plane inter-lacement. However, 3D braided fabrics have low transverse properties due to the absence of yarns being equivalent to the filling yarns in 3D woven fabric. They also have size and thickness limitations. Various methods and techniques have been developed for 3D braiding and these 3D braiding techniques are commercially available. Furthermore, various unit cell–based models for 3D braiding were developed to define the geometrical and mechanical properties of 3D braided structures. Most of the unit cell–based models include micromechanics and numerical techniques. Multiaxis 3D braided fabrics have multiple layers and show no delamination, and their in-plane properties are enhanced due to the ±bias yarn layers. However, the multiaxis 3D braiding technique is at an early stage of development and needs to be fully automated. This will be a future technological challenge in the area of multiaxis 3D braiding (Bilisik 2013). The two main factors likely to continue to influence research and development in the transport industry for the foreseeable future are the environment and the control of cost. The 2D and 3D braiding technologies, can contribute to preserving the environment by introducing lighter weight fabrics and devising ways to facilitate recycling of car components. Textile recycling poses a challenge because the textile face fabric is usually inseparably joined to another material, which is generally chemically dissimilar. The use of textiles to replace this material to help reduce the number of material types and hence facilitate recycling, presents opportunities. However, the most significant way in which the braided textile structures are likely to contribute to a better environment is to use composite materials that replace heavier metals and significantly reduce the weight of road vehicles. New high-performance materials that are being developed by thermoplastic fiber and braided preform structures present further opportunities for innovation in both products and production techniques. In addition, although textiles are already used in most areas of the transportation industry, it has been suggested that opportunities still exist for further usage such as in dashboards, door and seat pockets, seat backs, and sun visors. A new generation of fibers produced by biological synthesis is likely to appear in the future for better braided fabrics including higher standards of cleanability, antimicrobial finishes/antimicrobial fibers, better thermal comfort, antistatic properties, and less odors (Fung and Hardcastle 2001). Textile technology proves to be the most cost effective and efficient technology in providing structural parts with flexibility and high performance to support future development of transportation industries. These developments will result in light weight, clean, safe, and energy-efficient products.

REFERENCES

Akiyama, Y., Z. Maekawa, H. Hamada, A. Yokoyama, and Y. Uratani. 1995. Braid and braiding method. US Patent No. 5385077.

Algor Inc. 1997. ALGOR User's Manual. Pittsburgh, PA.

Belgrano, G. and C. O'Connell. 1992. *Carbon Sails to the Front*. TTI, May, 28–31.

Beyer, S., S. Schmidt, F. Maidi, R. Meistring, M. Bouchez, and P. Peres. 2006. Advanced composite materials for current and future propulsion and industrial applications. *Adv. Sci. Tech.* 50: 171–178.

Bilisik, A. 1991. Three dimensional (3D) weaving and braiding. PhD dissertation, University of Leeds, United Kingdom.

Bilisik, A. K. 1998. Multiaxial and multilayered 8-step circular braided preform for composite application. *8. International Machine Design and Production Conference*, Middle East Technical University, Ankara, Turkey.

Bilisik, K. 2011. Three dimensional (3D) axial braided preforms: Experimental determination of effects of structure-process parameters on unit cell. *Text. Res. J.* 81:2095–2116.

Bilisik, K. 2013. Three dimensional braiding for composites: A review. *Text. Res. J.* 83:1414–1436.

Bilisik, K. and N. Sahbaz. 2012. Structure-unit cell base approach on three dimensional (3D) representative braided preforms from 4-step braiding: Experimental determination of effect of structure-process parameters on predetermined yarn path. *Text. Res. J.* 82:220–241.

Brandt, J., K. Drechsler, and J. Filsinger. 2001. Advanced textile technologies for the cost effective manufacturing of high performance composites. *RTO AVT Specialist Meeting on Low Cost Composite Structures*, RTO-MP-069(II), Norway.

Branscomb, D., D. Beale, and R. Broughton. 2013. New directions in braiding. *J. Eng. Fibers Fabr.* 8:11–24.

Brookstein, D. S., J. Skelton, J. R. Dent, R. W. Dent, and D. J. Rose. 1994. Solid braid structure. US Patent No. 5357839.

Brown, R. T. 1988. Braiding apparatus. UK Patent No. 2205861 A.

Brunnschweiler, D. 1953. Braids and braiding. *J. Text. Inst.* 44: 666–686.

Buckley, J. D. and D. D. Edie. 1993. *Carbon-Carbon Materials and Composites*. New Jersey: Noyes Publications.

Bullock, K. R. and J. R. Pierson. 1992. *Kirk Othmer Encyclopedia of Chemical Technology*, 4th ed. New York: John Wiley.

Chen, J. L. and A. El-Shiekh. 1994. Construction and geometry of 6 step braided preforms for composites. *39th International SAMPE Symposium*, Anaheim, CA.

Chou, T. W. 1992. *Microstructural Design of Fiber Composites*. Cambridge, England: Cambridge University Press.

Cristaldi, G., A. Latteri, G. Recca, and G. Cicala, 2010. Composites based on natural fibre fabric. In *Woven Fabric Engineering*, P. D. Dubrovski (editor), 318–342. Croatia: Sciyo Publishing.

Donnet, J. B. and R. C. Bansal. 1990. *Carbon Fibers*. New York: Marcel Dekker.

Dow, M. B. and H. B. Dexter. 1997. Development of stitched, braided and woven composite structures in the ACT Program and at Langley Research Center (1985 to 1997). NASA/TP-97-206234.

Drechsler, K. 1999. Chapter XI: 3-D textile reinforced composites for the transportation industry 3-D textile reinforcements in composite materials. In *Composites Science and Engineering*, A. Miravete (editor), 10: 43–66. Cambridge, England: Woodhead Publishing Series.

Faust, H. S., E. M. Hogan, R. N. Margasahayam, and J. Hess. 1988. Development of an integral composite drive shaft and coupling. *National Technical Specialists Meeting on Advanced Rotorcraft Structures*, Williamsburg, Virginia.

Fiber Innovations Inc. 2002. Technical documents.

Florentine, R. A. 1982. Apparatus for weaving a three dimensional article. US Patent No. 4312261.

Fung, W. and M. Hardcastle. 2001. *Textiles in Automotive Engineering*. Cambridge, England: Woodhead Publishing and Technomic Publishing.

Gries, T., J. Stueve, T. Grundmann, and D. Veit. 2008. Textile structures for load-bearing applications in automobiles. In *Textile Advandces in the Automotive Industry*, R. Shishoo (editor), 301–319. Cambridge, England: Woodhead Publishing and CRC Press.

Hamada, H., A. Fujita, A. Nakai, A. Yokoyama, and T. Uozumi. 1995. New fabrication system for thick composites: A multireciprocal braiding system. *Innovative Processing and Characterization of Composite Materials* ASME, 20: 295–304.

Head, A. A. 2000. High coverage area braiding material for braiding structures. US Patent No. 6112634.

Hearle, J. W. S. 2001. *High-Performance Fibers*. Cambridge, England: Woodhead Publishing.

http://china-hdmi-cable.en.made-inchina.com/product/yovmSBXKPgcs/China-HDMI-to-HDMICable-HC030-.html. Accessed on March 25, 2014.

http://en.wikipedia.org/wiki/Rope. Accessed on March 17, 2014.

http://en.wikipedia.org/wiki/Cable. Accessed on March 20, 2014.

http://nhmarine.com.au/rope/double-braid-yacht-rope-polyester.html. Accessed on October 1, 2014.

http://sicksport.com/rock-climbing-climbing-rope-c-48_78/climbing-rope-lm-dynamic-ropes-p-1097. Accessed on October 1, 2014.

http://taflab.berkeley.edu/ME168-FA13/ME168_Applications.htm. Accessed on May 15, 2014.

http://www.braider.com. Accessed on September 20, 2014.

http://www.compositesworld.com/articles/autoclave-quality-outside-the-autoclave. Accessed on October 2, 2014.

http://www.compositesworld.com/articles/composite-engine-valves. Accessed on October 2, 2014.

http://www.directindustry.com/prod/tempo-international/hydraulic-hoses-rubber-wire-braided-78801-743861.html. Accessed on September 20, 2014.

http://www.pelicanrope.com/static_kermantle_roeps.html. Accessed on October 1, 2014.

http://www.ropeinc.com/climbing-safety-dynamic-ropes.html. Accessed on October 1, 2014.

http://www.nauticexpo.com/prod/baltic-ropes-magistr/yacht-dock-lines-polyester-core-polyester-sheath-33266-244907.html. Accessed on October 1, 2014.

http://www.wanderingtrail.com/Mods/JKmods/JKAiROCK/AiROCKTANKLINES.html. Accessed on September 20, 2014.

Janssen, H. 1996. Aramide fibres and new adhesive systems to elastomers, applications and performance. *6th Annual Conference of Textile Coating and Laminating*, Dusseldorf, Germany.

Kamiya, R., B. A. Cheeseman, P. Popper, and T. W. Chou. 2000. Some recent advances in the fabrication and design of three dimensional textile preforms: A review. *Compos. Sci. Technol.* 60: 33–47.

Ko, F. K. 1987. Braiding. In *Engineered Materials Handbook, Volume 1 Composites*, 519–528. Ohio, Canada: ASM International.

Ko, F. K. 2008. Textile composites for automotive structural components. In *Textile Advances in the Automotive Industry*, R. Shishoo (editor), 320–343. Cambridge, England: Woodhead Publishing and CRC Press.

Ko, F. K. and T. W. Chou. 1989. *Textile Structural Composites*. New York: Elsevier.

Kostar, T. D. and T. W. Chou. 1994a. Microstructural design of advanced multistep three dimensional braided performs. *J. Compos. Mater.* 28: 1180–1201.

Kostar, T. D. and T. W. Chou. 1994b. Process simulation and fabrication of advanced multistep 3-dimensional braided performs. *J. Mater. Sci.* 29: 2159–2167.

Kurbak, A. and A. S. Soydan. 2010. Small diameter circular textile materials and their manufacturing methods. *J. Text. Eng.* 17: 30–37.

Lambillotte, B. D. 1989. Fabric reinforcement for rubber. *J. Coated Fabr.* 18: 162–179.

Langer, H., A. Pickett, B. Obolenski, H. Schneider, M. Schneider, and E. Jacobs. 2000. Computer controlled automated manufacture of 3D braids for composite. *Euromat Symposium*, Munich, Germany.

Lee, S. M. 1990. *International Encyclopedia of Composites*, 130–147. New York: VHC Publisher.

Linden, D. 1995. Batteries and fuel cells. In *Electronic Engineers Handbook*, D. G. Fink and Christiansen (editors). New York: McGraw-Hill Education.

McConnell, R. F. and P. Popper. 1988. Complex shaped braided structures. US Patent No. 4719837.

McKenna, H. A., J. W. S. Hearle, and N. O'Hear. 2004. *Handbook of Fiber Rope Technology*. Cambridge, England: Woodhead Publishing and CRC Press.

Meriam, J. L. and L. G. Kraige. 1978. *Engineering Mechanics Volume 1, Statics*. New Jersey: John Wiley.

Mukhopadhyay, S. K. and J. F. Partridge. 1999. Automotive Textiles. *Textile Progress* 29: 97–107.

Mungalov, D. and A. Bogdanovich. 2004. Complex shape 3-D braided composite preforms: Structural shapes for marine and aerospace. *SAMPE J.* 40: 7–21.

Popper, P. 1990. Braiding. In *International Encyclopedia of Composites*, S. M. Lee (editor), 130–147. New York: VCH Publishers.

Popper, P. and R. A. McConnell. 1987. New 3D braid for integrated parts manufacture and improved delamination resistance—the 2-step process. *Proceedings of 32nd International SAMPE Symposium and Exhibition,* Anaheim, CA, United States.

Rogers, C. W. and S. R. Crist. 1997. Braided preform for composite bodies. US Patent No. 5619903.

Scott, J. R. 1971. Testing procedures and standards in rubber technology and manufacture. In *Rubber Technology and Manufacture,* C. M. Blow (editor), 446–477. London: Newnes-Butterworth.

Skelton, J. 1989. Triaxially braided materials for composites. In *Textile Structural Composites,* F. K. Ko and T. W. Chou (editors), 123–132. New York: Elsevier.

Strorage, W. and J. Ganter. 2014. *Physics for Cavers: Rope, Loads, and Energy.* http://nervenet.zocalo.net/jg/c/pubs/Rlenergy/Ropesloads.htm. Accessed on May 8, 2014.

Summerscales, J. 1987. Marine applications of composites. In *Engineering Materials Handbook, Vol. 1,* 837–844. Metal Park, Ohio: ASM.

Tsuzuki, M., M. Kimbara, K. Fukuta, and A. Machii. 1991. Three dimensional fabric woven by interlacing threads with rotor driven carriers. US Patent No. 5067525.

Uozumi, T. 1995. Braid structure body. US Patent No. 5438904.

Uozumi, T., Y. Iwahori, S. Iwasawa, and T. Yamamoto. 2001. Braiding technologies for airplane applications using RTM process. *Processing of the Seventh Japan International SAMPE Symposium,* 697–700, Tokyo, Japan.

Wallenberger, F. T. 1999. Structural silicate and silica glass fibers. In *Advanced Inorganic Fibers Processes, Structures, Properties, Applications*, F. T. Wallenberger (editor), 129–168. New York: Kluwer Academic Publishers.

Wallenberger, F. T., J. C. Watson, and H. Li, 2001. *Glass Fibers.* Ohio: ASM Handbook Composites.

Wootton, D. B. 2001. *The Application of Textiles in Rubber*. 187–212. Exeter, United Kingdom: Rapra Technology.

www.diytrade.com/china/4/products/3037377/Braided. Accessed on March 20, 2014.

www.fao.org/docrep/007/ad416e/ad416e06.htm. Accessed on May 5, 2014.

www.offshoreenergytoday.com/emerson-to-upgrade-safety-and-automation-systems-on-visund-platform. Accessed on May 6, 2014.

Yamamoto, T. and T. Hirokawa. 1990. Advanced joint of 3D composite materials for space structure. *35th International SAMPE Symposium*, Anaheim, CA, United States.

10 Conclusions and Future Directions

Sohel Rana and Raul Fangueiro

CONTENTS

ABSTRACT

This chapter presents a brief summary of the different subjects discussed in various chapters of this book and also presents the major conclusions. Some future directions for research and development in the area of braided structures and composites are also highlighted.

10.1 SUMMARY AND CONCLUSIONS

Due to the cost-effectiveness and capability of producing complex shapes, braiding has come up as a popular fabric manufacturing technique for various advanced technical applications. Three-dimensional (3D) fabric structures produced from braiding technique are preferred for applications in aerospace, transportation, civil infrastructure, and medical fields. Braided structures are usually applied in these various sectors in the form of composite materials. Braided composites can be produced in complex geometries and near to net shapes and, therefore, costs and wastes associated with the processes for obtaining desired shapes can be avoided. Additionally, braided composites present excellent mechanical performances such as high transverse strength and modulus, shear performance, damage tolerance and fatigue life, notch insensitivity, and fracture toughness. These interesting features combined with cost-effectiveness and huge flexibility in produced shapes make braided structures and composites highly suitable for advanced technical applications.

Among the various braided structures, biaxial and triaxial two-dimensional (2D) braided structures have already been considerably applied in different technical sectors and the techniques to produce these structures are well developed. Compared to 2D braided structures, 3D braided structures contain multiple layers and exhibit high delamination resistance due to out-of-plane interlacements and are highly suitable for producing complex geometries. However, they present lower in-plane mechanical properties as most braided tows remain off-axis from the loading direction and are highly crimped. Multiaxial 3D braided structures have improved in-plane mechanical properties due to the ±bias yarn layers. Different methods to produce 3D braided structures are also commercially available.

Braided composites can be produced using conventional composite manufacturing processes used for other types of composite materials. Resin transfer molding can be used to produce braided composites with very low defects in a cost-effective way. Braided rods can be produced directly using braiding technique having provision for resin impregnation, consolidation, and curing in a single step or combining braiding process with a pultrusion line. For producing thermoplastic braided composites, comingled yarns containing both reinforcing and matrix filaments can be braided to produce a hybrid perform, which is then passed through a pultrusion die to melt and consolidate the thermoplastic matrix yarns. The quality of the produced braided composites strongly depends on the temperature profile of the pultrusion die.

The properties of braided structures and composites are highly sensitive to process, material, and structural parameters. Analytical modeling of the braiding process parameters is performed to understand the relationship between the process parameters and the properties of braids. This approach helps to control the braiding process to produce braids with desired performance. Regarding the structural parameters, usually, the in-plane mechanical properties are mainly dependent on the type of fibers and the braiding angle. Lowering the braiding angle usually improves the in-plane mechanical performance. Mechanical performances can be further improved using axial yarns, and in these braided structures the mechanical properties are primarily dependent on the properties of axial yarns. Proper selection and hybridization of axial yarns result in excellent stiffness, strength, and ductility. The shear performance of braided composites is generally higher compared to conventional laminates and highly dependent on the braided component and the braiding angle. Braided composites also present very good energy absorption capability and crashworthiness, which are mainly controlled by fiber and matrix properties, fiber–matrix interface, and braiding angle. Compared to other composites, braided composites present a textured surface, which can be tailored easily through the manipulation of braiding process parameters, and this textured surface can play an important role in the bonding behavior of braided structures or composites with other types of matrices.

The structure and mechanical behaviors of 2D and 3D braided preforms can be successfully predicted with the help of microstructural modeling strategies from both braided preforms and unit cells. These modeling strategies can help in understanding the structure and properties of braided preforms for applications in composite materials. Mechanical properties of braided composites can be predicted through the finite element method using representative unit cell (RUC). In this approach, the effective properties of braided composites are calculated imposing periodic boundary conditions on RUC. To predict the failure and damage of braided composites, failure stresses and damage mechanisms of constituents, that is, braiding yarn and matrix, are considered and the failure criterion can be established using progressive damage analysis models. However, this modeling approach assumes the periodic characteristics of the mesostructure of the braided composites, and this may not be realistic for thin-walled braided composites with complex shapes. Some other problems also exist in this modeling approach related to the in situ mechanical properties of constituents, determination of initial damage criterion, and so on, and these issues have to be solved for better prediction of mechanical properties. Additionally, analytical techniques for other important mechanical properties such as fatigue, impact, and so on, have to be developed in the future.

Regarding the application of braided structures and composites, braided structures have been widely explored in medical fields for various applications including bidirectional barbed sutures, smart sutures using shape memory polymers, braided scaffolds for tissue engineered artificial cruciate ligaments, braided–pultruded rods for orthopedics, and so on. In civil engineering, braided composite rods are used for concrete reinforcement as a lightweight and durable replacement for steel rebars. Through the optimization of composition and structure, smart features such as self-sensing of strain and damage are also introduced in these braided rebars. Masonry wall strengthening is another application of braided structures in civil engineering. Meshes produced with braided structures have been developed to provide masonry walls sufficient resistance against damages due to earthquakes. Moreover, auxetic property has been introduced in these meshes to provide better strength and energy absorption capacity. Use of geogrids made from braided structures for soil stabilization is another application that is still in the development stage. Braided composites have huge potential for application in aerospace structures. Braided composites can be applied in rocket nozzle, fan blade containment case, aircraft propellers, stator vanes, and so on. Some applications that have already been commercialized are braided fan cases of GEnx turbofan engines, Honeywell stator vanes, Airbus A350 fuselage frames, and so on. Compressed braided gas tanks to replace filament wound gas tanks are also being developed for commercial applications. Braided structures are also utilized extensively in the transportation sector. Some applications of braided materials in the transportation sector include ropes, cables, drive shafts, structural elements, battery separators, hoses, belts, and so on.

10.2 FUTURE DIRECTIONS

From the various chapters of this book, it is clear that braided structures and composites possess a number of beneficial features that make them attractive for application in various technical sectors. A great deal of research and developments are being undertaken to overcome the current limitations of braiding technology, understand the structure and properties of existing products through experimental and analytical approaches, and also design new and better products for existing or new applications. Research and development efforts are helping in the commercialization of existing technologies and utilization of the products in practical applications. Besides improvement of the existing products, another strategy to extend the application of braided products should be the implementation of entirely new concepts to develop products with entirely new sets of properties.

Nowadays, nanotechnology is being widely applied as a means to develop multifunctional materials. In the composite sector, the use of nanotechnology and nanomaterials has been extensively investigated and various multifunctional nanocomposites have been developed with commercial success. With respect to braided composites, some efforts have already been observed to develop multiscale (combination of macro- and nanoscale materials) braided composites with self-sensing capability using carbon nanotubes. However, there is ample scope to develop different types of braided products utilizing different nanomaterials, as illustrated in Figure 10.1. Combinations of these various properties can be introduced within a braided structure or composite by incorporating various nanomaterials within the fibers or in the composite matrix or in the form of coatings. The multiscale material development approach may be much easier and effective to achieve the targeted properties compared to the conventional approach of modifying the existing structure or combining other macroscale materials. For example, a 2D braided composite laminate with dispersed nanotubes within the matrix can provide similar delamination resistance as 3D braided composites. Also, a multiscale braided composite may provide much better elastic modulus, strength, and ductility compared to hybrid braided composites containing a mixture of macroscale reinforcements. Additionally, some special features such as self-sensing capability, conductivity, electromagnetic shielding, gas barrier properties, and so on, are much more easier to achieve by combining multifunctional nanomaterials such as nanotubes.

However, there exist some technical challenges associated with the use of various nanomaterials, such as nanomaterial cost; toxicity; and processing problems such as agglomeration, and so on, and, therefore, further research is necessary to overcome these problems. Another critical issue in the development of multiscale materials is the lack of analytical techniques that can help in material designing and understanding of structure and properties. Analytical techniques for complex braided structures are still in the development stage. Therefore, combining another material from the nanoscale and analysis of the structure/properties of resulting materials are big challenges for researchers and scientists. Nevertheless, it is quite clear that these materials have a very good future due to their outstanding features and it is expected that a number of commercial products developed based on this concept will come up in the near future.

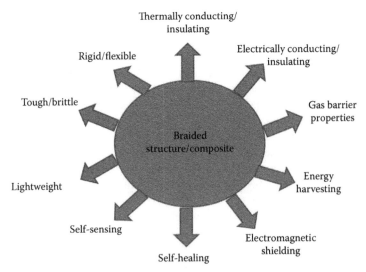

FIGURE 10.1 Concept of multiscale, multifunctional braided structures and composites.

Index

Printed and bound by CPI Group (UK) Ltd, Croydon, CR0 4YY
22/10/2024
01777613-0010